Power Electronics Applied to Industrial
Systems and Transports 5

Series Editor
Bernard Multon

Power Electronics Applied to Industrial Systems and Transports

---

Volume 5
Measurement Circuits,
Safeguards and Energy Storage

Nicolas Patin

ELSEVIER

First published 2016 in Great Britain and the United States by ISTE Press Ltd and Elsevier Ltd

ISTE Press Ltd
27-37 St George's Road
London SW19 4EU
UK

www.iste.co.uk

Elsevier Ltd
The Boulevard, Langford Lane
Kidlington, Oxford, OX5 1GB
UK

www.elsevier.com

Notices

Knowledge and best practice in this field are constantly changing. As new research and experience broaden our understanding, changes in research methods, professional practices, or medical treatment may become necessary.

Practitioners and researchers must always rely on their own experience and knowledge in evaluating and using any information, methods, compounds, or experiments described herein. In using such information or methods they should be mindful of their own safety and the safety of others, including parties for whom they have a professional responsibility.

To the fullest extent of the law, neither the Publisher nor the authors, contributors, or editors, assume any liability for any injury and/or damage to persons or property as a matter of products liability, negligence or otherwise, or from any use or operation of any methods, products, instructions, or ideas contained in the material herein.

For information on all our publications visit our website at http://store.elsevier.com/

British Library Cataloguing-in-Publication Data
A CIP record for this book is available from the British Library
Library of Congress Cataloging in Publication Data
A catalog record for this book is available from the Library of Congress
ISBN 978-1-78548-033-1

Printed and bound in the UK and US

Contents

Preface

This volume could have been called *"All which was not discussed in the first four books but which should not be overlooked!"*. Indeed, this book completes the four preceding books by addressing topics that, even though not an integral part of the power electronics field, are crucial to designing reliable and efficient converters. The first chapter deals with current, voltage, and temperature measurements. Such measurements are recurring issues in any power electronic converter. Components and sensors are introduced and specified by means of their technology, key characteristics, and conditioning circuits required to process information (whether by analog or digital circuits). The second chapter deals with shielding components used commonly in power electronics as well as in other electronic fields (especially in measuring devices). Indeed, the task at hand here is to study how "fragile" components can be shielded from damage caused by voltage, or intensity, surges that can occur in either faultless or malfunctioning circuits. Standards set for electronic protective setups are also mentioned. Additionally, part of the chapter focuses on explosion-proof equipment (ATEX) and on waterproof ratings of electrical devices (Protection Index concept – IP code). The final chapter deals with electrical energy storing elements commonly used in association with power electronic converters. This chapter addresses capacitors and supercapacitors first, then deals with batteries and accumulators. Part of the chapter focuses on the features of these components, especially monitoring, recharging, and establishing charge equilibrium (supercapacitors or batteries). The concept of accuracy is addressed in the first chapter which deals with measurements. Appendix 1, dedicated to uncertainty calculations, gives further information on how the performance of a measurement chain is characterized. As a final note, a short appendix (Appendix 2) introduces helpful information for converting between metric and imperial units.

Nicolas PATIN
January 2016

Sensors for Power Electronics

1.1. Overview

Regardless of the quantities that have to be measured, the sensors used in power electronics are characterized by generic parameters as well as in any other electronic field:

– precision;

– bandwidth;

– differential input (or single-ended input when the signal is taken relative to ground);

– galvanic insulation.

This last parameter is particularly important in power electronics in order to:

– maintain existing galvanic insulation of the power electronic converter for the sensor (refer to the Insulated Switching Power Supply section in Volume 3 [PAT 15c]);

– avoid propagating faults/parasitic effects from the power circuit to the control circuit (especially in the case of high power applications with high voltages and strong currents).

The unique characteristics of power electronics are due to the "chopped up" nature of measurable quantities (or at least of certain quantities in the system). This can cause measuring issues in terms of the bandwidth required for sensors that are not all suited to pick signals up presenting steep wavefronts (large $\frac{dV}{dt}$ and/or $\frac{dI}{dt}$). Furthermore, although theoretically the measured quantities do not suffer discontinuities (current through an inductance, voltage across a capacitor), their readings can be affected by other quantities undergoing large fluctuations (current or voltage through a transistor). Those fluctuations are likely to induce parasitic signals via capacitive or inductive coupling in the measuring circuit connectors of a printed circuit board, for example.

Broadly speaking, the routing of a printed circuit board for a power electronic converter requires paying careful attention to the following critical elements:

– power circuit connections;

– measuring circuit connections (to avoid introducing parasitic effects in the readings as much as possible);

– connections of control signals of electronic switches (to avoid oscillations/parasites, therefore, avoid unwanted switch commutations or at the very least avoiding slow commutations leading to additional losses).

In this respect, circuit miniaturization limits inductive coupling in the best way possible (and, to a lesser extent, capacitive coupling). It is always possible, using fine qualitative reasoning, to evaluate whether a given routing "geometry" is adequate or not. This qualitative analysis can even be carried out before performing a quantitative analysis (usually lengthy) requiring expensive and complex equipment (such as retrieval software for parasitic components due to component wiring, and simulation software). In order to conduct this reasoning, it is essential to focus only on the critical connections listed previously, and to neglect non-critical connections (power supply tracks for control circuits, if the latter are properly uncoupled, thanks to capacitors connected as close as possible to their power supply pins).

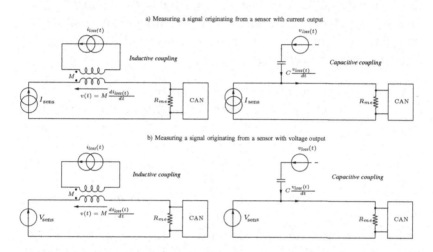

Figure 1.1. *Parasitic coupling schematics (inductive and capacitive) for sensors with current output a) and voltage output b)*

Moreover, knowing the nature of the quantities carrying the information at the sensor output allows us to determine the output signal coupling vulnerabilities (capacitive or inductive). For example, consider the case of a sensor supplying a *current* reflecting the measured quantity (this is the case for a large number of Hall effect sensors). Then, inducing an EMF on a long track separating the sensor from the data processing circuit will, in this case, have little impact on the measured signal. This is illustrated in Figure 1.1(a). Indeed, let us assume the sensor is an ideal current source and the signal measured at the acquisition circuit (for instance, an analog-to-digital converter – ADC) is obtained by recording the voltage across a resistor placed as close as possible to the ADC. Then under these assumptions, the sensor output impedance is a current source impedance i.e. theoretically infinite impedance (even though only very large in practice), therefore it will not be possible for the EMF induced by inductive coupling on the (possibly long) connecting tracks to inject a parasitic current in the loop. On the other hand, a capacitive coupling with a closed track undergoing strong $\frac{dV}{dt}$ would be able to inject a parasitic $C\frac{dV}{dt}$ type current, therefore affecting the acquisition circuit and, consequently, the readings.

On the contrary, the circuit connecting a sensor delivering a *voltage* (reflecting the measured quantity) to an acquisition circuit (Figure 1.1(b)) will not be vulnerable to capacitive coupling as long as the impedance of this circuit is low (which is the case for a sensor acting like an ideal voltage source). Consequently, if the circuit does not have too high an impedance (especially, a reactance due to track inductance), the $C\frac{dV}{dt}$ type current will be absorbed by the voltage source, and will not have any impact on the voltage measured by the ADC. However, the same circuit will be very vulnerable to inductive coupling. This is because inductive coupling will induce an EMF that will add to the voltage supplied by the sensor, resulting in the voltage perceived by the ADC.

It is important to keep in mind that capacitive and inductive coupling, although different in nature, are often due to bad routing (of the printed circuit board). Common denominators are:

– long parallel tracks carrying chopped up current and/or voltage for the power circuit, on the one hand, and the signal coming from a sensor and leading to an acquisition circuit (analog or digital) on the other;

– "forward" and "return" conductors placed too far apart that create an inductive loop (self-inductance) and introduce a vulnerability to induction from another circuit (mutual inductance).

Although proximity between the power circuit and the measuring circuit cannot be avoided (because of course, the sensor must measure a variable of the power circuit!), the problems listed above should be minimized as much as possible. Firstly, coupling can be minimized by tracing tracks as short as possible on the "power side" and on the "measuring side" and by angling them at 90° with respect to each other. Secondly, in the case of single-ended measurements, minimizing the distance between a "forward" and a "return" conductor can be facilitated by implementing earth planes. Otherwise, for differential measures, the problem can be solved by routing the two tracks in parallel with spacing as small as possible (even more efficient if these tracks are angled at 90° with respect to the power circuitery (or circuit) in order to minimize disturbances).

1.2. Current sensors

Common current sensors can be grouped in three broad families:

– shunts (resistors);

– current transformers;

– Hall effect sensors.

Each of these sensor categories have different performances in terms of accuracy and bandwidth (or more precisely, maximum measurable frequency). However, certain sensor types have the noteworthy ability to measure a direct current; for example shunts, whereas current transformers are grossly incapable of performing such a measurement.

1.2.1. *Current measuring shunts*

1.2.1.1. *Ohm's law and related complications*

Since a shunt operates like a resistor, its operation can simply be described using Ohm's law:

$$v = R.i \hspace{4cm} [1.1]$$

The voltage across the shunt is proportional to the current passing through it. Therefore, knowing the proportionality coefficient R is enough to be able to derive, from the voltage reading, the current circulating through the branch on which the shunt is connected(in series). Unfortunately, even assuming that the voltage was measured perfectly (which is never the case in practice), not knowing the value of R remains an issue. Before discussing this resistance measurement and calibration, let us recall the relation linking resistance R (in Ω) to the physical (resistivity ρ in Ω.m) and geometric (length l and section S in m and m^2) properties of the resistive element considered:

$$R = \rho \cdot \frac{l}{S} \hspace{4cm} [1.2]$$

In practice, a shunt should allow for the current to be measured without disturbing the power circuit excessively, through which the current circulates. In particular, this keeps the converter efficiency from being deteriorated to a greater extent as losses will inevitably be triggered by this component. Additionally, these losses might impact the readings themselves. Indeed, losses lead to heating and heating has an effect on material resistivity as well as on the geometric dimensions of the resistive element (thermal expansion). As a guideline, the resistivity at $300\,\mathrm{K}$ of some metals as well as their associated temperature coefficient α are exhibited in Table 1.1. The α coefficient corrects for material resistivity according to the following linear function:

$$\rho_{x\,K} = \rho_{300\,K} \cdot (1 + \alpha.(x - 300))$$ [1.3]

In practice, pure metals are not used to make resistors (precision resistors in in particular) but it is important to be aware that resistivity of the materials used varies with temperature (some with a positive temperature coefficient – PTC, while others with a negative temperature coefficient – NTC). This phenomenon is correlated to dimension variation caused by thermal expansion. More or less strong resistance variations should, therefore, be expected for components that were not designed specifically to obtain temperature stable resistive shunts (extreme cases are used as temperature sensors called PTC or NTC "thermistors" – refer to Chapter 2).

Metal	Resistivity (in Ω.m) @300 K	Temperature coefficient (in $10^{-3}K^{-1}$)
Silver	16×10^{-9}	3.85
Copper	17×10^{-9}	3.93
Aluminium	28×10^{-9}	4.03
Lead	208×10^{-9}	4.2
Nickel	87×10^{-9}	5.37
Iron	100×10^{-9}	6.5
Tungsten	59×10^{-9}	4.5

Table 1.1. *Resistivity and temperature coefficients of some metals*

1.2.1.2. *Seebeck effect*

Another phenomenon, known as the Seebeck effect, is at work in voltage measurements across a shunt. This effect is usually not critical in power electronic applications. The Seebeck effect is a thermoelectric effect that generates EMF at the interface between two metals when this interface is heated up to a certain temperature. Since the shunt is soldered to cables, or more probably to a printed circuit board, two interfaces can be identified (in reality, four due to solder seams involving an additional metal) and as many EMFs. The temperature component normally being uniform, voltages can be expected to compensate each other. In practice, this is not always the case, due to dimensions of tracks in particular. In these conditions, a temperature difference can appear and, as a consequence, a residual EMF remains (that is, the operating principle of thermocouples that are commonly used for measuring temperature in the industry). However, the order of magnitude of thermoelectrically generated voltages is so small that these voltages will normally be overshadowed by voltages induced effectively by current flow. It is often desirable to have a minimum of several dozens if not hundreds of millivolts across a shunt in order to get reliable and accurate current measurement, whereas thermoelectric effect generates voltages proportional to the temperature in Kelvins with (positive or negative) coefficients of the order of a few μV/K. A thermal imbalance of the order of a hundred Kelvins in a properly manufactured power electronic converter is not realistic. It is, therefore, impossible to introduce a "parasitic" voltage via this phenomenon, not even of the millivolt order. In contrast, this phenomenon is non-negligible while measuring very weak signals (voltage or current) and needs to be taken into account in the design of precision circuits (refer to publication [KEI 10] for further information on this topic). In order to overcome this problem, very weak shunts (for example $0,0001\Omega$ for a Vishay WSMS2908 shunt[1]) come equipped with connections using low thermal EMF alloys ($< 3\,\mu$V/°C). Potentials are established for each material and a list can be found on Wikipedia. On the other hand, with regards to connections with copper conductors (which are the most common on printed circuit boards), the EMF order of magnitude of a few couples are enough to evaluate the parasitic voltages introduced by this phenomenon (see

---

1 Those shunts are specialized for "wattmeter" type applications.

Table 1.2). When it comes to soldering seams, the $Cd_{70}Sn_{30}$ alloy can serve as a reference because its Seebeck coefficient with regards to copper is close to zero. This is not the case of common alloys such as the traditional standard lead $Sn_{60}Pb_{40}$ ($1 - 3\,\mu V/°C$) soldering, whereas an alloy with very high lead content also performs excellently ($Pb_{90}Sn_{10}$). It should, however, be noted that their use is problematic regarding regulations in force because of the presence of cadmium or lead, although exemptions from RoHS standards exist (lead-acid batteries for instance). In conclusion, the Seebeck effect has a weak impact on relatively high voltage measurements and it should not be considered a major problem for current measurements, as such, (even if the voltage across shunts is rather weak). The Seebeck phenomenon is even less of an issue for voltage measurements, unless dealing with very specific applications, for which, "manipulated" levels are weak and where measurement performances (in terms of accuracy) are high. It should, nevertheless, be noted that this type of effect can be compensated for in temperature measuring circuits with Pt100 type four-wire-connections probes where extreme accuracy is sometimes sought after (see section 1.4.3).

Metal couples	Seebeck coefficient Q_{AB} (in $\mu V/°C$)
Copper – Copper	$\leq 0.2\,\mu V/°C$
Copper – Silver	$0.3\,\mu V/°C$
Copper – Gold	$0.3\,\mu V/°C$
Copper – Lead/Tin	1 to $3\,\mu V/°C$
Copper – Silicon	$400\,\mu V/°C$
Copper – Kovar	~ 40 to $70\,\mu V/°C$
Copper – Copper oxide	$\sim 1000\,\mu V/°C$

Table 1.2. *Seebeck coefficients with respect to copper for some materials (source: [KEI 10])*

REMARK 1.1.– Component pins (passive or integrated circuits) are conventionally manufactured in Kovar. Kovar is an iron-based alloy comprising (in mass) of 29 % nickel, 17 %

cobalt and, in smaller quantities, carbon ($< 0,01\,\%$), silicon ($0,2\,\%$) and manganese ($0,3\,\%$).

1.2.1.3. *Aging*

When a resistor is functioning, it heats up. However, when heating up, the component starts ageing (as does any other electronic component). Before reaching an advanced state of decay, the resistance parameter is slowly altered. Ageing is inevitable and needs to be taken into account while designing a measuring circuit. Two different situations might arise:

– case 1: accuracy requirements are weak enough that they will not be affected by the decaying of the shunt resistor throughout the entire lifespan of the device;

– case 2: over time, the unmanaged resistance decay leads to an unacceptable deterioration of measurement accuracy.

The second case (not common in power electronics but frequent for measuring devices) will require the user to conduct recalibrations of the measuring circuit on a regular basis in order to compensate for the decay. In the first case on the other hand, this will not be necessary, or at the very least, not once the manufacturing process of the item has been completed (it is possible to perform an initial calibration in case the tolerance on the component value is insufficient for the accuracy expected).

As a guideline, the Vishay WSMS2908 shunt datasheet gives the following information on component decay throughout its lifespan: Load Life (1000 h at + 70°C, $1,5\,\mathrm{h}$ "ON", $0,5\,\mathrm{h}$ "OFF"): $\pm\,1,0\,\%$ ΔR.

1.2.1.4. *Resistor technologies and the concept of noise*

From the point of view of resistor technologies, let us focus, first of all, on power resistors: these are usually wound into a coil on a ceramic mount. They are designed to sustain important temperature rises and dissipate only a few watts.

They should not usually be used for current measuring applications in an electronic converter. The reason for this is that they have an inductive behavior that can cause problems if the shunt has to be placed in series with a switching component (however, they could be used if the shunt had to be placed in series with an inductor because, in a case such as this, the current would be a continuous function of time). Lower power resistors ($< 1\,\text{W}$) are built by layering carbon on a ceramic rod: those are the most commonly used through-hole resistors in analog electronics (especially during practical sessions on breadboards).

A matter, which has not yet been raised, concerns noise in resistors: free electrons are always in Brownian motion (erratic motion) as long as temperature is different from absolute zero ($0\,\text{K}$ or $-273.15°\text{C}$)[2]. This leads to a noise signal within the resistors that can be modeled by a randomly distributed voltage source with variance $\overline{v_b^2}$:

$$\overline{v_b^2} = 4k_B.T.R.B \qquad\qquad [1.4]$$

where k_B is the Boltzmann constant ($1.3806 \times 10^{-23}\,\text{J.K}^{-1}$), T is the temperature in K, R is the resistance in Ω and B is the bandwidth concerned in Hz. It should be noted that this expression of thermal noise (or Johnson noise) is well-verified for metallic resistors, but turns out to be too weak to account for noise in resistors with carbon layers: in this type of resistors, an additional $1/f$ noise adds to the white noise (noise whose power spectral density is constant for all frequencies) defined by equation [1.4]. These resistors with carbon layer share this characteristic with SMD type resistors (SMD stands for Surface Mounted Device) whose technology uses thick films (thick-film resistors). Those thick-film resistors are manufactured using a paste

---

2 Which is not always the case in practice!

containing a mixture of insulating elements and conducting elements (ruthenium oxide) deposited onto the surface of an alumina insulating substrate to a thickness of the order of several hundreds of μm. This technique makes it possible to obtain resistors with a standard accuracy of around 5 to 1% (at best 0.5%). These also have a limited temperature stability of several hundred ppm/°C (ppm = parts per million) or at best several dozen ppm/°C. Thick-film components can eventually be used as low-cost shunts for current measurements but will not give highly accurate results.

1.2.1.5. *SMD thin-film resistors*

For current measurement applications, SMD thin-film resistors turn out to perform considerably better (but are also more expensive). These resistors still rely on an alumina substrate, however their resistive element is obtained by depositing a thin nichrome (NiCr)[3] film (with a thickness of around one hundred nanometers). Since it is based on metallic film, noise is limited solely to Johnson noise (white noise). Furthermore, these resistors can be laser-trimmed efficiently (see Figure 1.2) in order to obtain accurate values (the best accuracy of off-the-shelf components is of the order of 0.01 %). Additionally, they are substantially temperature-insensitive (there are resistors that have a TCR[4] lower than 1 ppm/°C).

Last, but not least, important issue concerning SMD resistors (both thin and thick-film) is their frequency response: the absence of connection pins lowers the possibility of parasitic inductance of these components significantly. From there, current measurements obtained with surface-mounted shunts can be valid for a wider frequency range than with through-hole components. This can prove useful to design converters whose switching

---

3 As indicated by its name, it is a nickel and chromium alloy.
4 Temperature Coefficient of Resistance.

frequencies can exceed a megahertz (for resonance converters in particular).

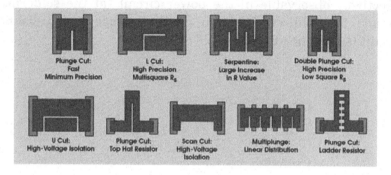

Figure 1.2. *Laser trimming motifs to adjust thin-film precision resistors (source: GSI Lumonics inc.)*

1.2.1.6. *Four-terminal shunt*

Connections between a shunt and the printed circuit board, on which it is placed, can introduce resistances that will impact and lower the measurement accuracy. Furthermore, these characteristics can disperse from one board to another. In these conditions, the accuracy of weak resistors can be compromised, thereby rendering them unusable. However, as was discussed previously, shunts should have a minimal impact on the circuits to which they are connected. Therefore, the use of low value resistors is crucial.

In order to get rid of this limitation, precision resistors (generally those that have a low resistance value) have not two but four terminals. This kind of packaging is explained by the fact that two of the terminals are dedicated to the current flow ("power" terminals) whereas the other two are used to pick up the voltage across the integrated resistive element. Since there is no current flowing through the voltage measurement circuit (or at the very least, a very low current because of the very high input impedance of the measurement circuit), no ohmic voltage drop can be triggered. As a consequence, even though voltage drops are effectively observable in the power circuit (at the soldering seams), they will not be "seen" by the voltage measurement circuit (see Figure 1.3).

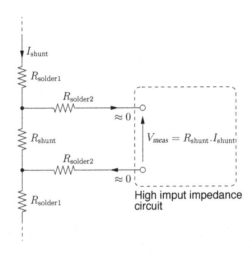

Figure 1.3. *Operating principle of a four-terminal shunt*

These types of components can be found in "through-hole" versions but also as SMDs as shown in Figure 1.4. In order to obtain the best accuracy possible, the voltage and current terminals should not be connected directly to shared pads on a printed circuit board: this is especially true for SMD shunts for which the "small protruding tabs" can, sometimes, go unnoticed. The solder pad blueprint shown in the figure is very clear: four pads (and not two!) have to be soldered onto the printed circuit board for the shunt to work properly. The larger pads are dedicated to current flow (of the power circuit) whereas the two smaller pads should be connected to the voltage measurement circuit (differential – see next section).

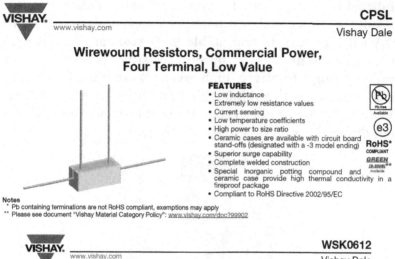

CPSL

Vishay Dale

www.vishay.com

Wirewound Resistors, Commercial Power, Four Terminal, Low Value

FEATURES
• Low inductance
• Extremely low resistance values
• Current sensing
• Low temperature coefficients
• High power to size ratio
• Ceramic cases are available with circuit board stand-offs (designated with a -3 model ending)
• Superior surge capability
• Complete welded construction
• Special inorganic potting compound and ceramic case provide high thermal conductivity in a fireproof package
• Compliant to RoHS Directive 2002/95/EC

RoHS*
COMPLIANT

GREEN
(R-2008)**
Available

Notes
* Pb containing terminations are not RoHS compliant, exemptions may apply
** Please see document "Vishay Material Category Policy": www.vishay.com/doc?99902

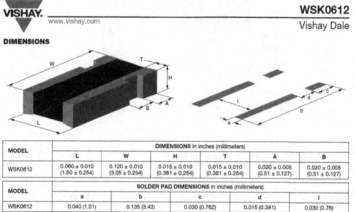

WSK0612

Vishay Dale

www.vishay.com

DIMENSIONS

MODEL	DIMENSIONS in inches (millimeters)					
	L	W	H	T	A	B
WSK0612	0.060 ± 0.010 (1.50 ± 0.254)	0.120 ± 0.010 (3.05 ± 0.254)	0.015 ± 0.010 (0.381 ± 0.254)	0.015 ± 0.010 (0.381 ± 0.254)	0.020 ± 0.005 (0.51 ± 0.127)	0.020 ± 0.005 (0.51 ± 0.127)

MODEL	SOLDER PAD DIMENSIONS in inches (millimeters)				
	a	b	c	d	l
WSK0612	0.040 (1.01)	0.135 (3.43)	0.030 (0.762)	0.015 (0.381)	0.030 (0.76)

Figure 1.4. *Four-terminal shunt casings (source: Vishay)*

Additionally, this type of shunt is not used exclusively for integrated measurements in an electronic device; it is also used in metrology. Figure 1.5 shows a picture of such a shunt from the brand LEM for which the measured voltage is picked up by a coaxial cable (there is a BNC meter socket at the bottom of the shunt baseplate). It is a very expensive component because of its accuracy and nominal range. Indeed, this shunt makes it possible to measure current up to 500 A with a accurate and stable 0.1 % resistance. As a side note, this shunt has a purely resistive behavior on a frequency range relatively high given its size (height: around 22 cm), with a current/voltage phase shift lower or equal than 0.1° up to 20 kHz (and a resistance that does not fluctuate by more than ±0.1 % on this frequency range).

Figure 1.5. *Power shunt from the brand LEM*

1.2.1.7. *Integration into a converter: differential or single-ended input*

Measuring a current using a shunt that has one of its terminals connected to the circuit ground, is the easiest, most comfortable situation to pick up, amplify, and eventually, filter the voltage across it in order to process the data by a control or monitoring device of a power electronic converter. Amplifying the voltage across the shunt by a gain suited for the downstream circuit is the most obvious solution to carry this measurement out. Otherwise, the voltage picked up can be sent directly to a dedicated terminal of the control/monitoring circuit of the converter. Figure 1.6 shows an example of a circuit board layout implemented for such a solution.

Typical Connection

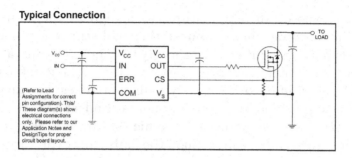

Figure 1.6. *Ground-referenced shunt used by a MOSFET driver (source: datasheet International rectifier IR2121)*

In the case where an op-amp (operational amplifier) is used to adapt the voltage across the shunt (usually weak) to the optimal processing range of the downstream circuit[5], care should be taken in order to maintain the expected current measurement accuracy. Thus, components should be chosen carefully to guarantee the same measurement quality as the shunt introduced in the power circuit. Constraints differ only in terms of temperature drift, provided that it is made sure that the temperature rise of this part of the circuit, caused by the shunt, will not spread to other parts of the device. Questions might eventually be raised concerning the required tolerance for resistors placed in an inverting or a non-inverting amplifier: measurement accuracy should be at least as good as the shunt accuracy if the device is intended to be operational in the absence of a calibration step. On the one hand, resistors can be of weak accuracy if a calibration is expected at the end of production line. On the other hand, these resistors must maintain a good stability over time. Additionally, it is without doubt preferable for these resistors to have a good temperature-dependent stability in order to guarantee a large functioning temperature range without loss of measurement accuracy. Even though these resistors

---

5 For example, an analog-to-digital converter input.

are not exposed to temperatures as a high as the shunt is (70°C for instance for the shunt), they will still be exposed to a warm environment of substantially fluctuating temperature during operations (at cold start or after a long intensive operation). Adding to that, they will, sometimes, be exposed to varyingly tempered environments (indoor or outdoor residential use, sometimes in vehicles). In the case of a system undergoing calibration, the only thing to verify is that, at worst, the assembly gain does not lead the acquisition stage (an ADC for example) to saturation under maximum current intensity.

Concerning the other elements of the analog circuit, selection criteria do not differ from conventional issues that arise when designing an op-amp circuit assembly:

– offset voltage ("auto-zero" type op-amp);

– symmetrical or asymmetrical power supply;

– supply voltage level;

– rail-to-rail output stage;

– bandwidth and slew-rate.

All these conventional issues remain true for voltage measurement across a floating shunt but in this case, the major issue is common mode rejection. Before discussing this problem and its existing solutions, it is worth discussing the reason why a floating shunt is very useful (see Figure 1.7). The most noteworthy advantage of using such a floating shunt configuration is that it allows us to maintain ground continuity between the input and output of a power electronic converter. Indeed, if an electronic card incorporates a shunt that introduces a potential difference between "input ground" and "output ground", it is strictly forbidden to connect these grounds together within the system where the card is integrated. If such a connection was indeed performed, the shunt would be short-circuited and all features associated with current measuring (regulation, limitation/protection) would then be lost. In the case where the shunt is placed on the "positive power line", ground continuity is secured. Input and output terminals can then be connected together (even if this is seldom used, it proves to be a good idea!). Therefore, this current measurement method can be considered more reliable when considering a "system approach" of said electronic feature.

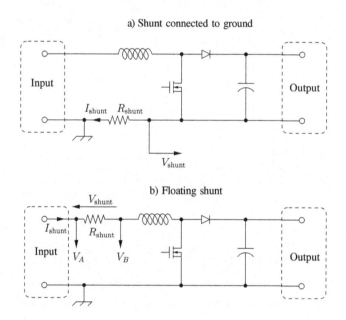

Figure 1.7. *Comparing different shunt configurations in a power electronic converter: a) shunt connnected to ground, b) floating shunt*

Now that the importance of this configuration type has been demonstrated, let us study the difficulties of implementing it in practice. For this, let us recall the goal that was initially defined: to measure current I_{shunt} flowing through the shunt by observing voltage V_{shunt} across it. Then, it becomes obvious that in order to carry out this measurement using a ground-referenced circuit assembly, two potentials have to be subtracted:

$$V_{\text{shunt}} = V_A - V_B \qquad\qquad [1.5]$$

In order to achieve this result, it is obvious that an op-amp subtractor circuit is the solution (see Figure 1.8). The following equation is obtained by applying the traditional approach for linear op-amp circuits ($V^+ = V^-$):

$$V_s = \frac{R_3\,(R_1 + R_2)}{R_1\,(R_3 + R_4)} \cdot V_2 - \frac{R_2}{R_1} \cdot V_1 \qquad\qquad [1.6]$$

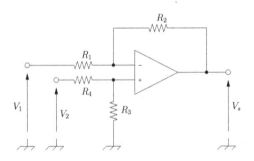

Figure 1.8. *Op-amp subtractor circuit schematic*

In theory, let $R_1 = R_4 = R_A$ and $R_2 = R_3 = R_B$, which leads to:

$$V_s = \frac{R_B}{R_A} \cdot (V_2 - V_1) \qquad\qquad [1.7]$$

and, in the particular case where $R_A = R_B$, this simply gives $V_s = V_2 - V_1$. Unfortunately, in practice, it is impossible to obtain strictly identical resistors. It is therefore necessary to analyze the actual result deterioration with regards to the theoretical case. In order to do this, let us assume the four resistors have approximately the same resistance value (nominal value which will be referred to as R_0). Then, the following simple equations are obtained:

$$\begin{cases} R_1 = R_0 + \Delta R_1 = R_0 \cdot \left(1 + \frac{\Delta R_1}{R_0}\right) \\ R_2 = R_0 + \Delta R_2 = R_0 \cdot \left(1 + \frac{\Delta R_2}{R_0}\right) \\ R_3 = R_0 + \Delta R_3 = R_0 \cdot \left(1 + \frac{\Delta R_3}{R_0}\right) \\ R_4 = R_0 + \Delta R_4 = R_0 \cdot \left(1 + \frac{\Delta R_4}{R_0}\right) \end{cases} \qquad [1.8]$$

Each expression of the resistances R_i (where $i \in \{1, 2, 3, 4\}$), has been rewritten to reveal parameter $\frac{\Delta R_i}{R_0}$ which, from now on, will be denoted α_i. The four α_i parameters are different but a given tolerance range normally imposes an upper bound on them: for example, for 0.1% resistors it is known that $|\alpha_i| \leq 0,001$. After replacing each of the R_i by their $R_0 \cdot (1 + \alpha_i)$ expression in equation [1.6], the following equation is obtained:

$$V_s = \frac{(1 + \alpha_3) \cdot (2 + \alpha_1 + \alpha_2)}{(1 + \alpha_1) \cdot (2 + \alpha_3 + \alpha_4)} \cdot V_2 - \frac{1 + \alpha_2}{1 + \alpha_1} \cdot V_1 \qquad [1.9]$$

Next, to continue with this study, it is possible to consider (reasonably) that for all i, $|\alpha_i| \ll 1$ and thus, that all products of the form $\alpha_p \cdot \alpha_q$ are negligible with respect to other terms of the developed equation [1.9]. The equations can be simplified as:

$$V_s \simeq \frac{2 + \alpha_1 + \alpha_2 + 2\alpha_3}{2 + \alpha_3 + \alpha_4 + 2\alpha_1} \cdot V_2 - \frac{1 + \alpha_2}{1 + \alpha_1} \cdot V_1 \qquad [1.10]$$

This equation is not very practical considering that the coefficients of V_1 and V_2 are asymmetrical. In order to obtain a more practical equation, the concepts of common mode amplification A_{mc} and differential mode A_d are introduced. V_s can, therefore, be rewritten as follows:

$$V_s = A_d \cdot (V_2 - V_1) + A_{mc} \cdot \frac{V_1 + V_2}{2} \qquad [1.11]$$

With this new equation, the subtracting or differential behavior (as expected) of the circuit clearly appears in the $V_2 - V_1$ terms. Whereas, the $\frac{V_1 + V_2}{2}$ term is the undesirable "common mode" term (it is, in fact, a parasitic term) which must be minimized as much as possible in practice (using a A_{mc} gain as low as possible). Next, it is possible to rewrite equation [1.11] in the same form as equation [1.10] by developing each term and regrouping the obtained V_2 and V_1 terms together:

$$V_s = \left(A_d + \frac{A_{mc}}{2} \right) \cdot V_2 - \left(A_d - \frac{A_{mc}}{2} \right) \cdot V_1 \qquad [1.12]$$

It then becomes easy to identify coefficients in equations [1.10] and [1.12] and to obtain the following expressions for A_d and A_{mc}:

$$A_d \simeq \frac{1}{2} \left(\frac{2 + \alpha_1 + \alpha_2 + 2\alpha_3}{2 + \alpha_3 + \alpha_4 + 2\alpha_1} + \frac{1 + \alpha_2}{1 + \alpha_1} \right) \qquad [1.13]$$

$$= \frac{1}{2} \left(\frac{(2 + \alpha_1 + \alpha_2 + 2\alpha_3) \cdot (1 + \alpha_1) + (1 + \alpha_2) \cdot (2 + \alpha_3 + \alpha_4 + 2\alpha_1)}{(2 + \alpha_3 + \alpha_4 + 2\alpha_1) \cdot (1 + \alpha_1)} \right)$$

$$A_{mc} \simeq \frac{2 + \alpha_1 + \alpha_2 + 2\alpha_3}{2 + \alpha_3 + \alpha_4 + 2\alpha_1} - \frac{1 + \alpha_2}{1 + \alpha_1}$$

$$= \frac{(2 + \alpha_1 + \alpha_2 + 2\alpha_3) \cdot (1 + \alpha_1) - (1 + \alpha_2) \cdot (2 + \alpha_3 + \alpha_4 + 2\alpha_1)}{(2 + \alpha_3 + \alpha_4 + 2\alpha_1) \cdot (1 + \alpha_1)}$$

Now that these two gains have been expressed (even though those are approximations in which all second order terms were neglected), the degree of perfection of a subtractor can be characterized by a quantity (expressed in decibels – dB) called the common mode rejection ratio (CMRR) which is defined as:

$$\text{CMRR}_{\text{dB}} = 20.\log\left|\frac{A_d}{A_{mc}}\right| \qquad [1.14]$$

The goal is, therefore, to obtain the highest CMRR possible in order for the common mode voltage to have the smallest impact possible on the subtractor output. From the A_d and A_{mc} expressions established previously, it is possible to derive an approximation for the CMRR depending on resistance uncertainty:

$$\text{CMRRdB} \simeq 20.\log \qquad [1.15]$$

$$\left|\frac{(2+\alpha1+\alpha2+2\alpha3).(1+\alpha1)+(1+\alpha2).(2+\alpha3+\alpha4+2\alpha1)}{2.((2+\alpha1+\alpha2+2\alpha3).(1+\alpha1)-(1+\alpha2).(2+\alpha3+\alpha4+2\alpha1))}\right|$$

To continue these calculations, it is necessary to neglect second order terms again in the developed numerator and denominator:

$$\text{CMRR}_{\text{dB}} \simeq 20.\log\left|\frac{4+5\alpha_1+3\alpha_2+3\alpha_3+\alpha_4}{2.(\alpha_1-\alpha_2+\alpha_3-\alpha_4)}\right| \qquad [1.16]$$

Here it can be seen that, since α_i coefficients are much smaller than 1 in absolute value, the numerator can be approximated by 4. In these conditions, the worst case scenario for the circuit CMRR consists of maximizing the denominator by letting:

$$\begin{cases} \alpha_1 = \alpha_{\max} \\ \alpha_2 = -\alpha_{\max} \\ \alpha_3 = \alpha_{\max} \\ \alpha_4 = -\alpha_{\max} \end{cases} \qquad [1.17]$$

In order to properly see the difficulty of obtaining a high CMRR, let us perform a numerical application by letting $\alpha_{\max} = 0,001$ (resistors with $0,1\%$ accuracy). This gives:

$$\mathrm{CMRR_{dB}} \simeq 20.\log\left|\frac{1}{2.\alpha_{\max}}\right| \simeq 54\mathrm{dB} \qquad [1.18]$$

Although this may seem to be high CMRR value (it corresponds to a A_{mc} gain 500 times smaller than A_d), the objective of this setup should be kept in mind: a voltage drop across a shunt, of the order of $1\,\mathrm{V}$ for example, has to be measured whereas the common mode voltage can reach $200\,\mathrm{V}$, or even more depending on the type of usage. In these conditions (assuming $A_d = 1$ and $A_{mc} = 1/500$):

$$V_s = 1 + \frac{200}{500} = 1,4 \qquad [1.19]$$

Therefore, with this setup, a 40% error on the output voltage of the subtractor arises, which of course, is unacceptable.

It is, however, not possible in reality to build a subtractor setup efficient enough for these types of applications using discrete components: it is necessary to use an integrated circuit of the "instrumentation amplifier" type which not only includes the op-amp, but also the resistors of the setup. This way, it is possible to match the resistors accurately (since they all have the same intrinsic value and are exposed to equal temperatures): this setup performs much better as can be seen on the AD8422 (*Analog Devices*) datasheet extract shown in Figure 1.9.

Admittedly these components are very efficient, however several issues should not be overlooked before choosing the adequate reference:

– the common mode rejection ratio depends on the gain (if it is an adjustable gain circuit such as the AD8422);

– the common mode rejection ratio deteriorates rapidly as frequency increases (see Figure 1.9);

– the maximum common mode voltage tolerated by the integrated circuit can be very weak (which precisely is the case of the AD8422 that has to be compared to the AD8479 – see Figure 1.10).

Parameter	Test Conditions/ Comments	AD8422ARZ			AD8422BRZ			Unit
		Min	Typ	Max	Min	Typ	Max	
COMMON-MODE REJECTION RATIO								
CMRR DC to 60 Hz with 1 kΩ Source Imbalance	$V_{CM} = -10$ V to $+10$ V							
G = 1		86			94			dB
G = 10		106			114			dB
G = 100		126			134			dB
G = 1000		146			150			dB
Over Temperature, G=1	$T = -40°C$ to $+85°C$	83			89			dB
CMRR at 10 kHz	$V_{CM} = -10$ V to $+10$ V							
G = 1		80			80			dB
G = 10		90			95			dB
G = 100		100			100			dB
G = 1000		100			100			dB

Figure 1.9. *Instrumentation amplifier AD8422 (Analog Devices) datasheet extract*

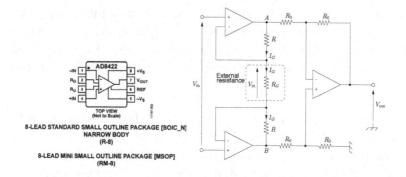

Figure 1.10. *Pin configuration diagram of the AD8422 (Analog Devices) and its internal layout schematic*

First, let us take an interest in the layout of a circuit such as the AD8422 amplifier to explain why the CMRR increases with gain. It can be seen in Figure 1.10 that this integrated circuit has two pins called "R_G" between which a resistor allowing us to adjust the gain is connected. The same figure shows on the right, the internal structure of the circuit leading to this result.

It can easily be verified that the common mode applied to the setup inputs is maintained (with a unit gain within the subtractor). Regarding the differential mode, the only point to note is that the voltage applied to the two input terminals can also be observed across resistor R_G. From there, the voltage applied across the two input terminals of the subtractor setup can be derived by applying (inversely) the voltage divider equation. Therefore, the following is obtained:

$$V_1 - V_2 = \left(1 + \frac{2R}{R_G}\right).V_{\text{in}} \qquad [1.20]$$

Since the global common mode gain is maintained whereas the differential mode gain increases (by a factor $1 + \frac{2R}{R_G}$, obviously superior or equal to 1), the common mode rejection ratio is, therefore, improved by the adjustment of R_G when the gain of the first stage becomes very high (in the datasheet, a maximum value of $150\,\text{dB}$ is indicated when the gain is 1.000).

As pointed out previously, the only performance related to common mode rejection should not overshadow other key characteristics of instrumentation amplifiers. The most important characteristic of all, for the purposes of power electronics, is the tolerated applied input voltage amplitude. Indeed, looking at the AD8422 datasheet it can be seen that input voltages cannot exceed supply voltages by more than $1\,\text{V}$. This clearly establishes an impediment which is often a deal breaker for power electronics. Using circuits such as the

AD8479 (even if it means a lower CMRR) that allow common mode voltages of $\pm 600\,\text{V}$ are preferred.

AD8422						Data Sheet
INPUT						
Input Impedance						
Differential		200‖2		200‖2		GΩ‖pF
Common Mode		200‖2		200‖2		GΩ‖pF
Input Operating Voltage Range¹	$V_S = \pm 2.3\,\text{V to} \pm 18\,\text{V}$	$-V_S + 1.2$	$+V_S - 1.2$	$-V_S + 1.2$	$+V_S - 1.2$	V
Over Temperature	$T = -40°\text{C to} +85°\text{C}$	$-V_S + 1.2$	$+V_S - 1.3$	$-V_S + 1.2$	$+V_S - 1.3$	V

Data Sheet						AD8479
INPUT						
Common-Mode Rejection Ratio (CMRR)	$V_{CM} = \pm 600\,\text{V dc}$					
	$T_A = 25°\text{C}$	80	90	90	96	dB
	$T_A = T_{MIN}$ to T_{MAX}	80		90		dB
	$V_{CM} = 1200\,\text{V p-p, dc to 12 kHz}$	80		80		dB
Operating Voltage Range	Common-mode		± 600		± 600	V
	Differential		± 14.7		± 14.7	V
Input Operating Impedance	Common-mode		500		500	kΩ
	Differential		2		2	MΩ

Figure 1.11. *Input operating voltage range comparaison for the AD8422 and the AD8479 (Analog Devices datasheet extract)*

Finally, the measurement of current circulating through a loop can be performed at different locations in a circuit (even if coupling the ground line has to be avoided, as discussed previously). One of the aims here is to minimize the common mode voltage fluctuations. Indeed, it was just pointed out that it is possible to find circuits suited to very high voltages. However, it is important to keep in mind that a high CMRR ($90\,\text{dB}$ for the AD8479) applies only to continuous components.

The CMRR drops quickly with frequency and can become insufficient if the common mode voltage is not only elevated but also (and especially in certain cases), if it varies at a high rate. To illustrate this, Figure 1.12 displays the CMRR plots of the AD8422 (on the left) and the AD8479 (on the right) over frequencies ranging up to $100\,\text{kHz}$ for one while $1\,\text{MHz}$ for the other.

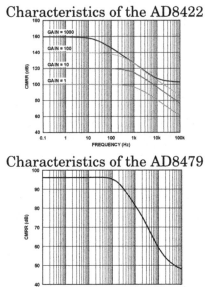

Figure 1.12. *CMRR as a function of frequency (source: Analog Devices). For a color version of the figure, see www.iste.co.uk/patin/power5.zip*

1.2.1.8. *Review: initial accuracy, temperature drifting, drifting over time and calibrations*

All the information provided, regarding current measurement using shunts (characteristics and physical phenomena that have an impact on readings and issues that arise due to the signal conditioning circuit) falls within the very general framework of measurements and assessment criteria for measuring processes. These measurement assessing criteria are as follows (they are illustrated in Figure 1.13):

– precision: indicates how large the standard deviation of the complete set of measurement data points obtained for a single value of the quantity measured is, meaning how the data is spread out around its central value. The standard deviation should be as small as possible for the process to be qualified as precise;

– trueness: indicates how close to the "true" value[6] the mean of the complete set of measurement data points is (obtained for a single value of the quantity measured). Of course, it is best for the statistical mean of the data to be true (that is, there is no bias between the data mean and the "true" value);

– accuracy: a sensor, or more precisely a measuring chain, is considered accurate if it has both excellent precision and trueness. In this case, measurements are always very close to the "true" value and it is not necessary to calculate the average over several data acquisitions to obtain a good result.

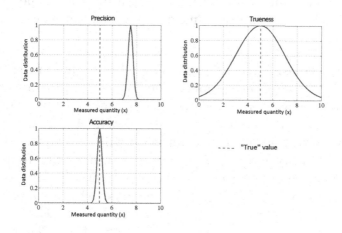

Figure 1.13. *Measurement precision, trueness and accuracy concepts. For a color version of the figure, see www.iste.co.uk/patin/power5.zip*

The desired outcome is, of course, the case illustrated in the third figure (accurate measurement case) with a small standard deviation, and no bias between the data mean and the "true" value. Generally speaking, this data bias is not an obstacle to obtaining a satisfying reading as long as the

---

6 In the hypothetical case that such a value exists!.

calibration is performed. There is still, however, a key constraint to performing this calibration (via software if the acquisition is carried out with an ADC): there needs to be a good reproducibility of this bias between runs no matter the operating mode. This corresponds perfectly to the precision constraint displayed in the first figure. Without this between-run precision, no calibration will allow an accurate measurement; generally speaking, the most sophisticated signal processing cannot counteract the deficiencies of a mediocre analog acquisition chain. Thus, the concept of noise, entirely or partially defined by its standard deviation[7], represents the limit below which a measurement cannot be improved by calibration. Noise amplitude in a measurement can only be reduced at the expense of a filter and, therefore, of an increase in response time of the measurement chain.

Now, let us review the theoretical elements discussed in previous sections regarding using shunts for measuring current. As long as a measurement is performed using a shunt whose resistor is stable over time and is temperature-independent, it will always be possible to perform a calibration. This calibration will, in turn, eradicate any uncertainty on the initial shunt resistance value from the acquisition chain (in case of a large tolerance on the shunt resistance value). Evidently, in the case where values are expected to drift over time, these shifts cannot be compensated for by the initial calibration. Therefore, recalibrations should be carried out on a regular basis in order to maintain a constant accuracy level over time. Finally, the most burdensome situation consists of setting up a measuring chain on which no calibration will be performed (neither initially, nor during the device lifespan). It is clear that, in this case, the accuracy of the initial resistance value

---

7 White Gaussian noise is in particular entirely defined by its standard deviation (also called "central moment of order 2" in signal processing).

will have to be raised (along with the temperature and time stability) in order to obtain reliable current measurements.

1.2.2. *Current transformer*

1.2.2.1. *Operating principle*

Measuring current using a shunt has several major advantages: accuracy, robustness, low cost (but it can be more expensive for high accuracy measures), and high bandwidth. This technique does, however, have a major downside: the absence of galvanic insulation between the power circuit[8] and the control/monitoring circuit in which the extracted signal is processed.

This difficulty can be overcome by setting up a device to separate the two circuits electrically. An efficient way to reach this goal, is to set up a magnetic coupling with, commonly, a toroidal core, as can be seen in Figure 1.14. In the case where the field is channeled by a zero-reluctance magnetic circuit, it can be written, in accordance with Ampère's law, as follows:

$$\sum_k n_k.i_k = 0 \qquad [1.21]$$

If the conductor, in which it is desired to measure the current i_p, only winds around the toroid once ($n_p = 1$), while the secondary winding (for measurements) is made of $n_s = n$ coils through which flow a current i_s, a relationship between currents is:

$$i_p + n.i_s = 0 \qquad [1.22]$$

and thus:

$$i_s = -\frac{i_p}{n} \qquad [1.23]$$

---

8 In which the targeted current flows.

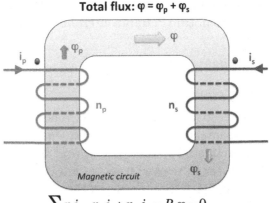

$$\sum n.i = n_1.i_1 + n_2.i_2 = R.\varphi = 0$$

- If the magnectic reluctance R of the circuit is assumed to be infinite

- If, by convention, currents are considered positive when entering through the homologous input points ● of each winding

Figure 1.14. *Diagram of the operating principle of a current transformer*

The current measured with an ammeter in the secondary winding is therefore proportional to the primary current and the gain of this structure is equal to $1/n$. This solution is therefore perfectly suited for measuring a strong current using a secondary circuit with enough coils for the current flowing through the measuring circuit to be reduced (for example with a maximum calibre of $10\,\text{A}$) while the primary current can be very high (for example $1\,000\,\text{A}$). This highlights a key parameter for transformers: their transformation ratio. Nevertheless, we are interested in the i_p/i_s ratio instead of the voltage ratio (secondary voltage/primary voltage) here, the same as in conventional transformer uses.

Moreover, it should be noted that the configuration of this transformer type is inverted as compared to standard transformers: the number of secondary coils is much higher than the number of primary coils. This, in fact, is a voltage step-up transformer and it must be used with a weakly resistive load in the secondary winding (typically an ammeter that behaves, if not as a short-circuit, then at least as a weak resistor). Therefore, a shunt such as those introduced in the previous section will naturally be used for a setup incorporating a current transformer. On the contrary, the secondary of a current transformer should never be left in open circuit as this would trigger a significant voltage surge (dangerous for the user and the system itself).

1.2.2.2. Technologies

Two conventional current transformer technologies can be distinguished (Hall effect sensors that are dealt with in section 1.2.3 set aside):

– "conventional" magnetic core transformers;

– iron-free transformers called *Rogowski coils*;

– Néel effect magnetic core transformers.

In the case of magnetic core transformers, the problem of material saturation arises, as well as the problem of iron losses that introduce phase shifts and, generally speaking, measurement errors. There exists different accuracy classes for current transformers and hence they can be categorized into the following two families depending upon the task at hand:

– some are in charge of measuring a nominal current (but also eventually a little beyond the nominal value);

– others have to operate as a protection (or shield) and must be capable of measuring currents way beyond the equipment nominal value[9].

---

9 And in this case, the issue of core saturation is of the utmost importance.

Generally speaking, ultra-soft materials (in other words, having a hysteresis cycle as thin as possible) showing the smallest possible eddy current losses will be selected[10].

On the other hand, it is inevitable for a magnetic material with a high μ_r to reach an excitation level H leading to a magnetic saturation beyond which there is no longer a linearity between H and B (the slope $\frac{dB}{dH}$ of property B(H) again tends to the magnetic permeability of the free space μ_0).

The advantage of Rogowski coils (which can also be considered as a downside) is that this device does not contain any ferromagnetic material to channel the magnetic field generated by the measured current. Rogowski coils are, consequently, capable of functioning at a very high frequency[11]. According to Faraday's law, the voltage v_e, developed at the output of the Rogowski winding, can be expressed as:

$$v_e = -\frac{nS\mu_0}{2\pi r} \cdot \frac{di(t)}{dt} \qquad [1.24]$$

where parameters S and r are the section of an elementary coil and the radius of the total loop respectively in accordance with the schematic representation in Figure 1.15, n is the number of coils and μ_0 is the magnetic permeability of free space (which is equal to $4\pi 10^{-7}$ V.s/(A.m)).

The intrinsic gain is weak and it requires an integrator circuit (active[12]) to recover a voltage reflecting the measured current. The most common solution consists of using an

---

10 A hysteresis cycle exists in low frequencies and widens due to eddy currents as frequency increases (refer to the iron loss Bertotti model).

11 Within the boundaries of the downstream integrator circuit.

12 Thus requiring a power supply contrary to magnetic core current transformers.

op-amp based setup for this purpose. Nevertheless, since the integrator is used in open loop, the presence of an offset at the op-amp input would, in practice, render this setup unusable if it was purely integrated[13]. In order to make this setup usable, it is therefore, necessary to limit the low frequency gain by adding a resistor (R_1) in parallel with capacitor C. The resulting transfer function is:

$$H_{\text{true}}(p) = \frac{v_s(p)}{v_e(p)} = \frac{-R_1}{R_0} \cdot \frac{1}{1 + R_1 C p} \qquad [1.25]$$

otherwise, with just the integrator (and no R_1 resistor):

$$H_{\text{int}}(p) = \frac{-1}{R_0 C p} \qquad [1.26]$$

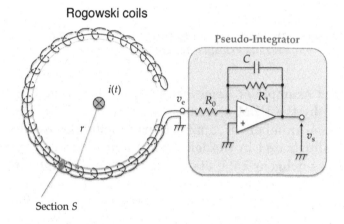

Figure 1.15. *Rogowski coils setup and its op-amp based pseudo-integrator*

13 The output voltage would rapidly drift towards one of the supply voltages (and therefore lead the setup to saturation) depending on the sign of the op-amp offset voltage.

As a consequence, the setup behaves like a pure $-R_1/R_0$ gain for angular frequencies significantly lower than $\omega_c = \frac{1}{R_1 C}$, but behaves as an integrator (as expected) with transfer function $\frac{-1}{R_0 C'p}$ for angular frequencies significantly higher than ω_c. In order to minimize the residual offset at the integrator output, it is preferable to use an op-amp with a very weak offset voltage (for example, an "auto-zero" type amplifier).

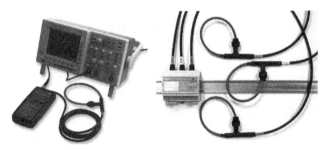

Figure 1.16. *Rogowski probes for oscilloscopes (on the left) and for industrial systems (on the right)*

A good example of Rogowski setup is the range of probes designed by the PEM company[14] (see Figure 1.16). They are used in a number of applications, including oscilloscope probes (illustrated in the left figure) and industrial systems such as switchgear cabinets (illustrated in the right figure). This type of sensor is very promising for carrying out accurate measurements with frequencies significantly exceeding a megahertz (16 MHz for PEM CWT probes). However, this is not a recent concept. It was developed in the first half of the 20th century (in 1912 to be precise) by W. Rogowski and W. Steinhaus [ROG 12].

Both of the two technologies discussed previously (magnetic core current transformers and Rogowski coils) have a limited ability to perform measurements in lower

---

14 Power Electronics Measurements Ltd, www.pemuk.com.

frequencies. Under no circumstances can they perform measurements of direct currents. There exists, however, a similar technology (without measuring the magnetic field directly) that makes these kinds of measurements possible: Néel effect sensors were named after Louis Néel who, in 1948, discovered the existence of materials exhibiting a non-linear magnetization (superparamagnetism) of the form:

$$M(H) = \chi_0 H + N_e.H^3 + \varepsilon\left(H^3\right) \qquad [1.27]$$

where χ_0 is the magnetic susceptibility in zero field and N_e is the Néel coefficient of the material. The $\varepsilon\left(H^3\right)$ term is evidence of a third-order Taylor series (polynomial approximation) of the magnetization curve. This approximation turns out to be accurate enough on the range of H excitation values actually used in practice. The third-order term draws interest because, in presence of a pulsed current (thus possible Fourier series decomposition of this current under periodic conditions), terms containing products between harmonics and the direct component of the excitation will appear. Therefore, the direct component of the current will be indirectly measurable at a measurement winding.

To verify this assumption, let us recall that induction B can be expressed as a function of H in the following way:

$$\mathbf{B} = \mu_0\mu_r\mathbf{H} \qquad [1.28]$$

which is, in fact, an adapted form (that introduces relative permeability of the magnetic material) of the free space relation:

$$\mathbf{B} = \mu_0\mathbf{H} \qquad [1.29]$$

Another way to write equation [1.28] invlolves the introduction of a material reaction term (magnetization M):

$$B = \mu_0 \left(H + M \right) \tag{1.30}$$

In the case of a conventional unsaturated material (for which $N_e = 0$), this gives:

$$B = \mu_0 \left(1 + \chi_0 \right) . H \tag{1.31}$$

and it is now possible to establish a link, for a given material, between χ_0 and relative permeability μ_r. In the case of a material used in a Néel effect sensor, the third-order term makes the material more complex to model (with two characterizing parameters instead of one). Nevertheless, this reveals a rather interesting behavior. Admitting that excitation $H(t)$ (from now on, equations involve moduli instead of vectors) can be expressed in the following form:

$$H(t) = H_0 + H_1 \cos \left(\omega t \right) \tag{1.32}$$

As a result, the following induction $B(t)$ expression is obtained:

$$
\begin{aligned}
B\left(t \right) = {} & \mu_0 \left(1 + \chi_0 \right) \left(H_0 + H_1 \cos \left(\omega t \right) \right) \\
& + \mu_0 N_e \left(H_0 + H_1 \cos \left(\omega t \right) \right)^3
\end{aligned}
\tag{1.33}
$$

From this expression, a constant term can be extracted (constant terms are not measurable with a coil, because, according to Faraday's law, an EMF can only be extracted at the coil terminals from the variable terms of induction). Terms with different angular frequencies values can also be extracted. All this is because:

$$
\begin{aligned}
\left(H_0 + H_1 \cos \left(\omega t \right) \right)^3 = {} & H_0^3 + 3 H_0^2 H_1 \cos \left(\omega t \right) \\
& + 3 H_0 H_1^2 \cos^2 \left(\omega t \right) + H_1^3 \cos^3 \left(\omega t \right)
\end{aligned}
\tag{1.34}
$$

and also because:

$$\begin{cases} \cos^2 x = \frac{1}{2}\left(1 + \cos 2x\right) \\ \cos^3 x = \frac{3\cos x}{4} + \frac{\cos 3x}{4} \end{cases} \qquad [1.35]$$

As a consequence, several components can be found:

– a component of angular frequency ω and of amplitude $\mu_0\left(1 + \chi_0\right)H_1 + \mu_0 N_e \left(3H_0^2 H_1 + \frac{3H_1^3}{4}\right)$;

– a component of angular frequency 2ω and of amplitude $3\mu_0 N_e H_0 H_1^2$;

– a component of angular frequency 3ω and of amplitude $\mu_0 N_e H_1^3 \cos^3\left(\omega t\right)$.

For the purposes of the Néel effect, the coil is not purely passively used: it is injected with a current that generates an excitation H_{exc} which overlaps the external field H_{ext} (generated by the current subject to measurement). The two fields will then interact (or "mix"), in the same way as the two decomposition terms of the unique field in the previous equation did. These two decomposition terms can be considered as two distinct fields generated by two separate sources: one constant and one sinusoidal. Therefore, the development can be written as:

$$\left(H_{\text{exc}} + H_{\text{ext}}\right)^3 = H_{\text{exc}}^3 + 3H_{\text{exc}}^2 H_{\text{ext}} + 3H_{\text{exc}} H_{\text{ext}}^2 + H_{\text{ext}}^3 \qquad [1.36]$$

Generally speaking, the "mixing" of the two field components makes the external field analysis possible around a carrier frequency (in a range of analysis frequencies in the manner of a spectral analyzer). In this instance, let us focus, in particular, on the $3H_{\text{exc}}^2 H_{\text{ext}}$ term. This term brings the H_{ext} field to around twice the excitation frequency f_{exc} of field H_{exc}. This technique is the basis of the measurement method used by Néel effect sensors developed by the Neelogy company (www.neelogy.com, see Figure 1.17). It can be seen from this figure that these sensors are very similar to Rogowski coils that are also available for purchase. The core of Neelogy sensors is made from a composite superparamagnetic nanostructured material.

Figure 1.17. *Néel effect current sensor (source: Neelogy)*

In fact, the actual device manufactured uses a negative feedback (counteraction) auxiliary winding whose purpose is to cancel out the second harmonic signal picked up at the excitation winding. These observations lead to finding that the current I_s in the feedback winding is equal (to the nearest number of coils N_s this winding counts) to the primary current I_p subjected to measurement (the one which created the initial external field H_{ext}):

$$I_s = I_p/N_s \qquad [1.37]$$

This measurement approach by negative feedback is actually also used in Hall effect sensors (at the very least in those operating in closed loop) which are the focus of the next section.

1.2.3. *Hall effect sensors*

1.2.3.1. *Operating principle*

The Hall effect was discovered in 1879 by Edwin Herbert Hall. It is characterized by the appearance of a voltage between two opposite faces of a parallelepiped conductive (or semi-conductive) bar. The induced voltage is oriented perpendicularly to the current flow in the bar and to the magnetic field in which the bar is immersed.

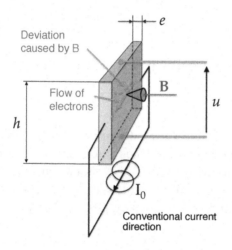

Figure 1.18. *Hall effect principle*

This is illustrated in Figure 1.18. If the current flowing through the bar is controlled, and considering the voltage is proportional to the magnetic field (a fact which was proven to be true in this section), this setup turns out to behave like a magnetic flux density sensor. By noting that moving electrons in a magnetic field are subject to a force which is perpendicular to their velocity vector, this "conventional"[15] Hall effect can easily be written into an equation:

$$\mathbf{F}_{mag} = q.\mathbf{v} \times \mathbf{B} \qquad [1.38]$$

This force leads electrons to divert from their natural trajectory (with $\mathbf{B} = 0$) and to accumulate on one of the bar faces. This generates an electric field that counteracts the effect of the magnetic field by a Coulomb force:

$$\mathbf{F}_{elec} = q.\mathbf{E} \qquad [1.39]$$

The equilibrium condition is therefore given by:

$$\mathbf{v} \times \mathbf{B} = -\mathbf{E} \qquad [1.40]$$

And it is then necessary to express this equality with measurable electrical quantities such as the current $i = I_0$ in the bar and the voltage u between the bar faces perpendicular to the field and the current flow. For this, simply recall that current density $\mathbf{j}$ in the bar is:

$$\mathbf{j} = n.q.\mathbf{v} \qquad [1.41]$$

where, n is the free charge (electrons or holes) density/concentration per unit volume. Furthermore, this current density is related to current through the section area $S_b = h.e$ (h: height, e: thickness) of the bar:

$$\|\mathbf{j}\| = \frac{I_0}{S_b} \qquad [1.42]$$

---

15 A quantum Hall effect also exists. As indicated by its name, it falls under quantum mechanics and is therefore more complex to model than the "conventional" Hall effect.

As for voltage u, it is related to the electric field (assumed to be uniform and vertical on the entire height h of the bar):

$$u = -\int_0^h \mathbf{E}(z).\mathbf{e}_z dz = -\mathbf{E} \cdot \mathbf{e}_z.h = -E.h \qquad [1.43]$$

Equation [1.40] can then be rewritten (admitting that $\mathbf{j}$ and $\mathbf{B}$ are orthogonal):

$$\frac{I_0.B}{nqS_b} = \frac{u}{h} \qquad [1.44]$$

where $B = \|\mathbf{B}\|$. It is possible to rewrite the equation as follows:

$$u = \frac{I_0}{n.q.e} \cdot B \qquad [1.45]$$

This proves that the voltage measured is proportional to the modulus B of the induction field (assumed perpendicular to the bar) as well as to the current I_0 flowing through the sensor. On the other hand, it is inversely proportional to the carrier charge (electrons in a metal or holes in a semi-conductor and so, in both cases $|q| = 1.6 \times 10^{-19}$ C), to the bar thickness e and to the charge density n per unit volume. This goes to show that there are only two parameters on which the sensor design should concentrate:

– sensor thickness should be minimized as much as possible;

– charge carriers concentration should also be minimized.

This last parameter leads to favoring semi-conductors over metals for the manufacturing of Hall sensors sensitive enough to be usable in practice. Indeed, semi-conductors are characterized by a low quantity of very mobile charges whereas metals possess about one free electron per atom (so 6.022×10^{23} e- per mole). However, the free electrons in metal have very low mobility (conduction speed is less than one millimeter per second, of the order of $0, 4$ mm/s to be more precise). It also should be specified that Hall sensors are not exclusively dedicated to current measuring: they are also used to measure electric actuators position (in particular in "DC Brushless" type motors) but also in position coders (for example, SKF instrumented rollers).

1.2.3.2. *Open loop measurements*

Different approaches can be adopted to measure a current flowing through a wire or a printed circuit board track. The first solution to achieve this measurement consists of using the Hall effect measurement of the field generated by the current to be measured directly. This type of sensor makes it possible to assess a direct current and an alternative current in the same way: therefore, it does not have the same limitations as current transformers. The major issues that arise when implementing this solution reside in the sensor gain and the measurement uncertainty in the absence of calibration. Figure 1.19 displays an example of a component operating on this principle.

Figure 1.19. *LEM FHS 40-P/SP600 sensor (evaluation kit no. 7)*

This is a sensor of brand LEM in an SMD casing whose reference is FHS 40-P/SP600. Several routing configurations for printed circuit boards are suggested depending on the desired sensitivity for the current that has to flow through the power circuit (this current goes up to 100 A maximum in a one-track configuration). In fact, the manufacturer offers an evaluation kit (component on the top face and track on the lower face) characterized by a nominal current of 30 A with a maximal measurable current of 76 A (26 mV/A sensitivity). The kit displayed in Figure 1.19 has slightly better greater sensitivity. This is because the track in which the measured current flows is looped several times, and because this track is situated on the same side as the component. As a consequence, the field is more intense for any given current. However, this limits the maximal measurable current (10 A) as well as the Galvanic insulation. Indeed, it should be kept in mind that this type of component allows us to separate the measuring circuit from the power circuit electrically (as do current transformers). However, this means respecting isolation distances between tracks adapted to the voltage level used (refer to Volume 1, Chapter 6).

1.2.3.3. *Closed-loop measurements*

In order to improve the efficiency of measurements, the magnetic field can be channeled using a ferromagnetic circuit in which the Hall sensor is incorporated. This setup still has the ability to measure direct currents at very low frequency. However, some of the same problems, already present in current transformers, arise again:

– iron loss in the core (which degrades the high frequency performances of the sensor);

– non-linearity of the material (saturation).

The first problem can only be tackled by the use of adapted materials (ultra-soft materials with weak hysteresis and limiting Foucault currents thanks to sufficiently thin layering[16] or due to a high resistivity[17]). The second problem can be solved thanks to the implementation of a negative feedback winding (as for Néel effect sensors discussed in the previous section). To carry this solution out, the current simply has to be steered in this auxiliary winding in order to cancel out the magnetic field measured by the Hall sensor. This way, the operating point of the device is maintained in a quasi-linear area of the material. Next, the current flowing in this winding has to be measured and from there the primary current can be deduced. This is possible because a compensation of "ampere-turns" happens: $I_s = I_p/N_s$ (where N_s is the number of coils of the feedback winding)[18].

Many Hall sensors operate on this principle and are actually directly equipped with a current output. Therefore, connecting a resistor between the output and ground is enough to recover a voltage reflecting the current to be measured. These sensors require symmetrical voltage supply between $+15\,$V and $-15\,$V. And also, the recovered voltage can vary approximately between these two values (even if the output is not exactly of "rail-to-rail" type) when using a sensor capable of measuring positive or negative currents.

Nevertheless, it should be said that certain sensors directly supply an output voltage. Some of them, in particular, are adapted to interface with a microcontroller; they require a voltage supply between 0 and 5 V and their zero-current output is set at $2,5\,$V. Their output increases under a positive current or decreases under a negative current. These circuits are additionally equipped with a reference output $(2,5\,$V$)$ for connecting the sensor to the

---

16 In the case of conventional Fe-Si type sheet metal for example.

17 Iron powders, ferrites or other similar materials.

18 This result (the same as for the Néel effect sensor studied previously) is of course only true if the primary current I_p only circulates once in the sensor. Otherwise, the number of primary coils N_p has to be included in the relationship which consequently becomes: $I_s = \frac{N_p I_p}{N_s}$.

differential input of an analog-to-digital converter if necessary. Finally, different types of sensor casing are available depending on their mounting characteristics: some have to be incorporated on a printed circuit board, others in a switchgear cabinet. In case of a mounting on a printed circuit board, the power circuit can be printed on the circuit board itself, or it can also be implemented using cables, coppers rods, or laminated busbars. As a result, current sensors come in all kinds of shapes, as portrayed in Figure 1.20.

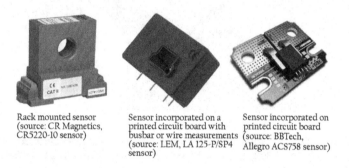

Rack mounted sensor (source: CR Magnetics, CR5220-10 sensor)

Sensor incorporated on a printed circuit board with busbar or wire measurements (source: LEM, LA 125-P/SP4 sensor)

Sensor incorporated on printed circuit board (source: BBTech, Allegro ACS758 sensor)

Figure 1.20. *Different Hall effect current sensors*

Oscilloscope current probes very often belong to this category. Two examples of these probes with $20\,A_{RMS}$ and $150\,A_{RMS}$ calibers respectively are displayed by Figure 1.21. It is crucial to establish the "zero" signal (when zero current is applied) of these probes to avoid a measurement bias: this calibration can be performed manually (using a potentiometer) or automatically by the probe. Lastly, even though it is not always the case, certain oscilloscope current probes are equipped with a *"Degauss"* feature. The purpose of degaussing is to demagnetize the core and therefore reduce, if not cancel, the remanent field B_r of the magnetic material (see Figure 1.22). This degaussing feature is well known by users of "old" computers. Indeed, the cathode ray tube (CRT) monitors of these older computers were also equipped with this feature. It was used in case the tube had been magnetized by an insufficiently magnetically shielded speaker (inducing deformations or, more often, coloured spots on images displayed by the CRT monitor). The degaussing feature may, for example, urge the material into excitation cycles aiming to bring the remanence back to a low level[19].

Figure 1.21. *Tektronix brand Hall effect current probes (TCP0020 on the left and TCP303 on the right) with* $20\,\mathrm{A_{RMS}}$ *and* $150\,\mathrm{A_{RMS}}$ *calibers respectively*

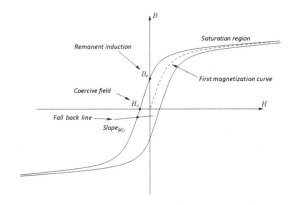

Figure 1.22. *Characteristic curve of a soft magnetic material*

---

19 In theory, a magnetized material follows a straight line (called a fall back line) of slope μ_0 in the $B(H)$ plane.

1.3. Voltage sensors

1.3.1. *Challenges associated with a straightforward function*

Voltage measurement can seem a lot simpler than current measurement because, in general, electronic set-ups, and in the ones based on op-amps in particular, voltages are manipulated more often and more proficiently than currents. Nevertheless, measured voltages are often high and are not always referenced with respect to ground. A good example of this is measuring the voltage across the different cells (connected in series) of a battery pack in an effort to manage the charge level (within a BMS – Battery Management System). The study of differential or instrumentation amplifiers previously carried out for voltage measurement across a shunt will not be carried on here because the challenges are identical and so are the solutions (taking into account the CMRR and the common mode voltage that the amplifier has to withstand). Galvanic insulation and ways to obtain galvanic insulation, however, will be discussed in this section (as for current transformers and Hall effect current sensors).

1.3.2. *Non-isolated measurement*

This section will deal with high-voltage measurements and the implementation of simple voltage divider bridges. This setup is probably one of the simplest one can face in electronics but, nevertheless, it has to perform a number of tasks:

– its primary function is to attenuate the voltage in order to allow a low voltage measuring circuit to function (an analog-to-digital converter);

– it has to show an impedance high enough to consume as little energy as possible and, therefore, to avoid degrading the efficiency of the power electronic converter;

– it should not overheat (to avoid resistor decay and premature ageing of components).

The ability of components to withstand voltage is therefore a crucial element when looking to largely miniaturize devices (particularly SMD devices). The concepts of clearance and creepage (introduced in Volume 1 [PAT 15a], Chapter 6) are therefore key elements that have to be taken into account when designing a divider bridge on a printed circuit board, in particular with surface mounted components. For instance, digging "trenches" in the printed circuit board to improve electric insulation is an interesting solution (refer to creepage). For voltages higher than 1 kV, the use of conventional insulators should also be avoided in favor of more adapted materials such as polyimide *Kapton* which is commonly used in "flex" type printed circuits [SIE 04]. Even the slightest contamination of the surface of insulators can lead to leakage currents that might considerably modify the value of resistors present on the printed circuit board. A simple calculation of the modification (induced by a fingerprint for example) of a $1\,G\Omega$ resistor when placed in parallel with a $100\,G\Omega$ resistor can serve as a guideline: the $1\,G\Omega$ resistor falls to around $990\,M\Omega$. This modification, although weak (1 %), can have an impact of measurements requiring high accuracy.

The operating principle of measurement probes (for voltmeters or oscilloscopes), such as the one portrayed in Figure 1.23, is to place a large value resistor encapsulated in a good quality insulator in series with the measurement circuit. Furthermore, there is, as in all passive probes, a compensating network to optimize the measurement bandwidth.

Given that power electronics are increasingly present in very high voltage networks, it can be useful be aware of the challenges inherent to these domains. The major challenge resides in insulator strength and in the issue of partial discharges: these phenomena can occur in solid insulators and lead to their progressive deterioration (see Figure 1.24). A phenomenon similar to discharges can also occur in liquids and gases and especially in the air: in this context, it is known as the corona effect (see Figure 1.25). Either way, the insulator is rendered conductive because the existing electric field becomes greater than the dielectric strength of the material. When this phenomenon is only local and transient, it is called a partial discharge. However, when this phenomenon is sustained over time between two conductors, it is called an electric arc. Finally, when it is on the scale of the Earth's atmosphere, it is called a lightning bolt!

Figure 1.23. *Testec brand high voltage probe HVP40 (40 kV DC)*

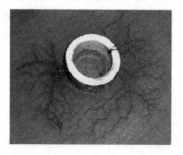

Figure 1.24. *Picture of an insulator which has been subjected to partial surface discharges (source: Wikipedia)*

Figure 1.25. *Picture of a corona discharge (source: Wikipedia)*

This insulation breakdown phenomenon is facilitated by the local expansion of the electric field. This expansion is induced by:

– small bending radii on the surface of a conductor (sharp edges, points, etc.);

– small suspended contaminants in the insulator (airborne conductive dust, air bubbles in a solid insulator etc.).

Theoretically, it is easy to verify the electric field expansion in the vicinity of a conductor with low bending radii. To achieve this, let us take the case of a charged sphere. Assuming this sphere of radius R carries a charge $+q$, it is possible to derive an expression for the radial electric field in spherical coordinates by directly applying the Gauss theorem:

$$\oiiint \mathbf{E}.d\mathbf{S} = \frac{Q_{\text{int}}}{\varepsilon_0} \qquad [1.46]$$

Through symmetry, the integral of flux **E** over the surface of a sphere of radius $r > R$ can be reduced to:

$$E(r).4\pi r^2 = \frac{q}{\varepsilon_0} \qquad [1.47]$$

where $E(r)$ is the modulus of $\mathbf{E}$ which is known to be radial (collinear with steering vector $\mathbf{e}_r$ in spherical coordinates). Therefore:

$$E(r) = \frac{q}{4\pi\varepsilon_0 r^2} \qquad\qquad [1.48]$$

It is possible to refer to potential V of the sphere and admitting that potential at infinity is zero, field $E(r)$ can be integrated on the interval $[R, +\infty]$:

$$V(R) - V(+\infty) = V(R) = \int_R^{+\infty} E(r).dr$$

$$= \frac{-q}{4\pi\varepsilon_0}\left[\frac{1}{r}\right]_R^{+\infty} = \frac{q}{4\pi\varepsilon_0 R} \qquad\qquad [1.49]$$

Therefore, a proportionality relationship has been established between the sphere potential and its charge: capacitance $C = 4\pi\varepsilon_0 R$ appears (in Farads, such that $q = C.V$). It is then possible to replace charge q with its expression as a function of voltage, allowing to rewrite the field generated by the sphere as a function of V:

$$E(r) = \frac{R.V}{r^2}, \ \forall r > R \qquad\qquad [1.50]$$

Thus, it can be seen that, for a given potential V, as R tends to zero, the electric field tends to infinity in the vicinity of the sphere. Generally speaking, any conductor having a bending radius approaching zero and subjected to a non-zero potential, will induce an intense local electric field concentration. This conductor can then easily cause a local breakdown of the insulator wrapping the conductor. In practice, conductors can never have a zero bending radius (even those of prismatic type that have sharp edges). Nevertheless, this problem becomes particularly critical in high voltage (also called HT for "high tension") applications and must be taken into consideration. On HT power lines, it can be seen (Figure 1.26) that particular solutions have been implemented: "guard rings" or corona rings (or anti-corona rings) as seen on Figure 1.26. The primary function of those rings is to encompass high voltage components (to which they are electrically connected) and to exhibit to the surrounding air an electric field as uniform as possible. Indeed, their toroidal-shaped geometry prevents the electric field from concentrating in certain regions where, without these components, it could reach levels critical to the dielectric strength of air (around $36\,\mathrm{kV/cm}$).

Figure 1.26. *Guard rings arranged on top of a high voltage (230 kV) power transmission line tower (source: Wikipedia)*

REMARK 1.2.– It can be seen on the probe handle shown in Figure 1.23 as well as on insulators visible in Figures 1.25 and 1.26 that no smooth elements are used for insulators. The reason for this is discharges frequently occur on the surface of insulators due to contaminations. Piling up discs on the surface lengthens the distance a potential discharge would have to travel. This workaround is therefore used very frequently for high voltage insulators.

1.3.3. *Galvanic insulation solutions*

Galvanic insulation is mandatory to ensure maximum protection of the control/monitoring circuit of a power electronic converter. How to achieve galvanic insulation for current measurements has been previously studied. However, the situation is more complex in the case of voltage measurements because there is no equivalent to the Faraday law in this case. This led to using a mirror-image of the current measuring shunt for the voltage measurement. Indeed, certain sensors (such as the LEM LV 25-P sensor – see Figure 1.27) use a resistor subjected to the measurable voltage, and they then measure the current flowing through that resistor using a Hall sensor. The galvanic insulation of such voltage sensor is therefore provided by a current sensor! This process is understandably called an *indirect voltage measurement*.

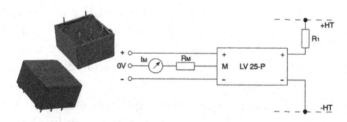

Figure 1.27. *Voltage measurement performed by a resistor associated with a Hall effect sensor (LEM LV 25-P)*

Now that the indirect approach to voltage measurement has been described, let us take a look at the direct approach. This approach relies on different modulation and insulation techniques to transmit information all the way to the control/monitoring circuit:

– insulation by capacitive barrier;

– insulation by piezoelectric transformer;

– insulation by magnetic transformer;

– insulation by optocoupler.

Most solutions implemented in industrial systems are essentially integrated setups that incorporate capacitive barriers, magnetic transformers and optocouplers. Nevertheless, specific setups incorporating piezoelectric transformers could very well be implemented to achieve a similar outcome:

– Texas Instruments manufactures a range of amplifiers insulated by a capacitive barrier: ISO121, 122, 124 (high precision, low cost), AMC 1100, 1200 and 1200-Q1 for automotive applications[20];

– analog devices manufactures magnetic transformer amplifiers: AD202, 203, 204, 208, 210 and 215;

– as for optical coupling, it is the speciality of the Avago company. It uses a Sigma-Delta modulation between a analog-to-digital converter (ADC) and a digital-to-analog converter (DAC) including components HCPL-7850 and 7851.

Low-tech solutions (using optocouplers operated in linear regime) can also be implemented. This is proven by the setup implemented in Chapter 5 of Volume 3. Nevertheless, these low-tech solutions often turn out to be slow (with a high response time), complicated to implement and lacking robustness (sensitive to temperature among other things)

Getting around the shortcomings of galvanic insulation for analog signals can however be advantageous as suggested by the Silicon Labs company with its digital couplers of the Si86xx family[21]. For the purpose of exploiting them in an

---

20 N.B.: AMC1x00 components are insulated amplifiers dedicated to measuring voltages across shunts (and thus to indirect measurement of currents)

21 These insulators operating on CMOS technology seem very similar to the Texas Instruments technology with the exception that they appear to be equipped with a double capacitive barrier (and a better withstand voltage characteristic).

analog context, a first stage should be used to decompose a continuous quantity into pulse signals (with modulated duty cycles). Then, at the coupler output, a low pass filter should be used to recover a signal reflecting the original signal.

In reality, the idea of a digital intermediary has to be considered in the current context, where most commands of electronic devices are delivered by digital components (microprocessors, microcontrollers, DSP, CPLD or FPGA). As a consequence, going back to the analog domain after passage through the galvanic insulation device can seem superfluous. With this observation in mind, different manufacturers offer isolated analog-to-digital converters:

– analog devices: AD7403, 7405 ADE7912;

– Texas instruments: AMC 1203, 1204, 1210 and AMC1304, 1305 dedicated to measuring voltages across shunts);

– Maxim: MAX78700 (dedicated to power measurements over two channels: voltage/current);

– Silicon Labs: Si8900, 8901, 8902.

Naturally, it is also possible to use conventional analog-to-digital converters galvanically linked to the power circuit on which the measurement is performed, before implementing a digital insulator such as those manufactured by Silicon Labs. Incidentally, it can be noted that solutions specifically dedicated to SPI buses (synchronous series link often implemented for analog-to-digital converters) are available: for example, Iso-SPI circuits manufactured by Linear Technology (LTC6820).

1.4. Temperature sensors

1.4.1. *Metrological challenges*

Temperature is a difficult quantity to accurately measure in the sense that thermal contact between the sensor and the element whose temperature is to be measured is almost hardly ever perfect. The only temperature a sensor is capable of measuring with very good accuracy is its own, nothing else. However, if there is no heat source in the sensor[22], the sensor temperature will probably be very close to the temperature of the element in contact with the sensor. This only applies if the element predominantly cools down through its surface in contact with the sensor. This hypothesis is generally well verified seeing as sensors are small in size compared to power components that are to be instrumented. Nevertheless, the major issue resides in the thermal inertia of elements separating semi-conductor chips (which are the heat sources) from the temperature sensor. Indeed, when interested in the temperature of a power device, the latter has to be placed on a heat sink and the temperature sensor has to be mounted on this same heat sink (as close to the power device as possible). The sensor will then be able to sense the temperature dynamic of the entire "device baseplate/heat sink" assembly which is, in fact, a filtered image[23] of the chip temperature. This kind of data cannot actually be useful to thermally protect the component since this type of instrumentation will probably introduce too great a detection delay compared to the required response time for component protection.

1.4.2. *Temperature sensor categories*

Temperature sensors can first be categorized depending on their working temperature range:

– NTC (negative temperature coefficient) type thermistors whose working range extends from -200 to $1,000°C$;

– PTC (positive temperature coefficient) type thermistors whose working range is generally limited to $0–100°C$;

---

22 Which can be considered true with great confidence.

23 Of "low pass" type.

– semi-conductor sensors whose working range is generally limited: for example for the KTY81 it extends from -55 to 150°C (source: NXP);

– thermocouples whose working range extends (depending on models – see Table 1.3) from -270 to $2\,600$°C[24];

– platinum probes Pt100 ($R_{0°C} = 100\Omega$) or Pt1000 ($R_{0°C} = 1\,000\Omega$) whose working range can extend from -50 to +500°C[25].

Family	Type	Metal couples	Temp. range (steady / intermittent state)
	E	Chromel / Constantan	0 °C - 800 °C / -40 °C - 900 °C
	J	Iron / Constantan	-20 °C - 700 °C / -180 °C - 750 °C
Base metals	K	Chromel / Alumel	0 °C - 1 100 °C / -180 °C - 1 200 °C
	N	Nicrosil / Nisil	0 °C - 1 150 °C / -270 °C - 1 280 °C
	T	Copper / Constantan (Copper + Nickel alloy)	-185 °C - 300 °C / -250 °C - 400 °C
	M	Nickel Molybdenum 18% (lead +) / Nickel Cobalt 0.8% (lead -)	+400 °C - +1 370 °C
	R	Platinum-Rhodium (13 %) / Platinum	0 °C - 1 600 °C / 0 °C - 1 700 °C
Noble metals	S	Platinum-Rhodium (10 %) / Platinum	0 °C - 1 550 °C / 0 °C - 1 700 °C
	B	Platinum-Rhodium (30 %) / Platinum-Rhodium (6 %)	100 °C - 1 600 °C / 0 °C - 1 800 °C
	C (or W5)	Tungsten-Rhenium (5 %) / Tungsten-Rhenium (26 %)	50 °C - 1 820 °C / 20 °C - 2 300 °C
Refractory met.	G (or W)	Tungsten / Tungsten-Rhenium (26 %)	20 °C - 2 320 °C / 0 °C - 2 600 °C
	D (or W3)	Tungsten-Rhenium (3 %) / Tungsten-Rhenium (25 %)	20 °C - 2 320 °C / 0 °C - 2 600 °C

Table 1.3. *Different categories of thermocouples*

---

24 One range of thermocouples does not cover the entirety of this temperature range.

25 This range can be narrower depending on the sensor.

The temperature range of thermocouples are often too wide for ordinary components demands (with the exception of T type probes). The same goes (to a lesser extent) for "platinum" (Pt) probes. However, they are often too expensive especially considering that their robustness to chemical attacks is not necessarily required in power electronics. Finally, NTC and PTC thermistors as well as semi-conductor probes are best suited for this type of application. Meanwhile, implementing sensors to get good measurements remains challenging:

– concerning ambient temperature measurements, semi-conductor sensors are best suited, both in a through hole casing and as a SMD[26] (see Figure 1.28);

– concerning temperature measurement on a heat sink, the following components can be used: either thermistors that can be fastened to the board by passing a screw through an eyelet perforated in the sensor body, or a sensor casing already equipped with a threaded stem (Figure 1.29) that can be fastened through a threaded opening in the heat sink of the controlled power component.

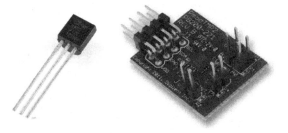

Figure 1.28. *Two semi-conductor sensors to mount on a printed circuit board (Analog Devices TMP36 on the left and ADT7420 on a Digilent PMOD TMP2 board on the right)*

---

26 N.B.: The sensor in a SMD casing of Figure 1.28 not only incorporates the analog sensor but also an entire conditioning circuit generating a 16bit digital signal (on a I^2C bus).

Figure 1.29. *Two types of thermistors to mount on a heat sink (source: EPCOS)*

"Platinum" probes, on the other hand, are used in certain energy storing components (supercapacitor packs for instance). With this in mind, it becomes clear that an adapted conditioning circuit is vital to benefit from the abilities of the probes. This will be discussed further in the next section.

1.4.3. *Examples of conditioning circuits*

This section deals with conditioning circuits used to recover temperature data with the help of a Pt100 type thermistor (RTD). In fact, different probe variants exists and their connectivity has a direct impact on the conditioning circuit they are paired with. Generally speaking, measurements are always carried out under a weak current intensity to limit the probes $R.I^2$ self-heating. Additionally, a calibration allowing for the extrapolation of the zero-current behavior is required to obtain accurate measurements (with no systematic bias). Under these conditions, excellent measurement accuracy is achievable (error lower than 0.1°C and certain "Pt100 probe + conditioner" acquisition systems such as datalogger PT104 (see Figure 1.30) guarantee a 0.01°C accuracy with a 0.001°C resolution).

A probe equipped with two connecting wires, can be inserted in a bridge (such as a Wheatstone bridge) with which a voltage unbalance V_{meas} will be analyzed (see schematic (a) of Figure 1.31). This configuration is the least accurate out of all the configurations displayed by this figure. This is because the connecting wires' resistance modifies the probe resistance. A calibration is still possible but this solution will inevitably be influenced by connections whose temperature is not under control. Furthermore, this simple solution (and as such generally uses a cheap conditioning circuit) would be unable to compensate for Seebeck effect induced EMF.

Figure 1.30. *Four-channel datalogger for Pt100 platinum probes (Pico PT-104)*

Generally speaking, when a greater accuracy is desired, either a three-wire or a four-wire connection will be used between the conditioning circuit and the probe. When a three-wire connection is used (Figure 1.31, schematic (b)), it becomes possible to compensate ohmic drops in connecting cables: this makes it possible to move the sensor even further away from the conditioning circuit (up to 600 m, whereas the two-wire solution is limited to 100 m). Finally, the most sophisticated solution (Figure 1.31, schematic (c)) uses a four-wire connection in which a current (of weak intensity) is injected in order to then measure the voltage across the temperature measuring platinum element. In these conditions, ohmic voltage drops in cables are perfectly compensated (since the voltage measuring circuit has a very high impedance). Additionally, from the setup symmetry, it is reasonable to assume that the Seebeck effect induced EMF is greatly compensated in twos. However it also possible to compensate for them by switching the flow direction of the injected current I. In these conditions, the measured voltage across the probe will be given by:

$$V_{\text{meas}} = \pm R_{\text{Pt}100}\left(\theta\right).I + V_{\text{Seebeck}} \qquad [1.51]$$

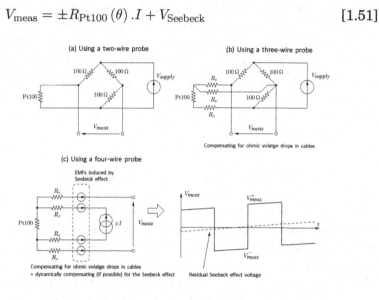

Figure 1.31. *Wiring diagrams of two, three or four-wire Pt100 probes*

This equation clearly proves that the sign of the term introduced by Ohm's law is modified by the current flow direction whereas the total voltage induced by Seebeck effect at every junction between different metals does not change sign. It is then possible to compare the voltages measured before and after the direction switch (respectively referred to as V_{mes}^{+} and V_{mes}^{-}) for the purpose of estimating the V_{Seebeck} term to better compensate for it. The EMF is not necessarily constant and it should be re-estimated on a regular basis to detect and counter a possible temperature drift. Nevertheless, the EMF fluctuations are usually slow compared to the feasible time scales of current switches with the conditioning circuit. Dynamic compensating is not an insurmountable problem provided that the settling period[27] of the measured voltage is quick enough.

In all three configurations (two, three or four-wire probe), the voltage measurement performed (in a bridge or directly across the probe) can advantageously benefit from an instrumentation amplifier (see section 1.2.1.7). This is because the amplifier helps rejecting common mode voltages fed at the input of the measuring circuit by electromagnetic disturbances. Those disturbances are in fact introduced because long cables are used to connect the active part of the probe to the signal conditioning circuit.

1.4.4. *Thermal protection of power components*

External sensors make it possible to manage to some extent the working temperature of components: they can for example be used to adjust a forced ventilation as they can give control over fan rotation speed. Nevertheless, the most efficient way to ensure component protection is to use a sensor directly incorporated in a power module. Not all modules are equipped with sensors but the majority commonly are. The technology of these sensors is generally similar to the technology used in PTC or NTC thermistors. An example of such a technology is shown in layout schematics of Semikron modules such as the one displayed in Figure 1.32. As a side note, the thermistor symbol used here is not standard (see Chapter 2) but it still easily understandable. The two arrows pointing up clearly indicate that when temperature goes up, the resistance also increases: this is therefore a PTC type thermistor. If it were a NTC thermistor, both arrows would point in opposite directions.

WARNING 1.1.– Even in with this integrated configuration, it is important to keep in mind that the measured temperature

---

27 Due to parasitic inductances and capacitances of the probe wiring.

is not that of the electronic components. This is clearly indicated in [SEM 14] where it is said that for electrical insulation reasons, the sensor has to be placed at an edge of the Direct Bonded Copper (DBC) baseplate. This means that the sensor gives a good measurement of the baseplate temperature but not of the chips that are soldered on top of the baseplate. Seeing as this configuration is not specific to a particular manufacturer, this finding can be extended to all power modules equipped with temperature sensors available on the market.

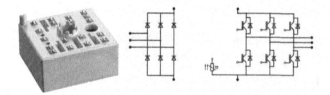

Figure 1.32. *Power module and its internal layout schematic (Semikron skiiP 01NAC066V3)*

Finally, one last thing should be pointed out concerning temperature components: there exists one other category of temperature related components called thermally triggered switches. Although they are not really capable of measuring temperature, they can ensure protection against overheating by themselves. They can perform such protection simply by triggering an alarm giving the order to cut off power when an excessive temperature is reached. Examples of such circuits in this category include the components of the ADT640x and ADT650x families (or the older TMP01) from Analog Devices. These components are not only capable of generating a signal warning when temperature exceeds a given upper threshold but also when temperature drops below a given lower limit (they are in fact natural candidates to the implementation of a thermostat – hence their name, thermostatic switches).

1.5. Measuring instruments

1.5.1. *Measurement selection and quality*

Measurement issues cannot be avoided at the laboratory level nor can they be avoided during control/maintenance operations at factory level. For every measurement that has to be carried out, several questions need answering before the most suitable device can be selected for the given objective. Some of these questions are similar to *in situ* measurement issues that arise for certain systems, these include:

– the desired accuracy;

– the value range of the measured quantity (depending on instrument ranges);

– the technology used by the device depending on usage constraints (digital or analog – meaning it has a pointer[28]),

– the device bandwidth (linked to measurement accuracy, and even to a possible deterioration if the working range is exceeded);

– verification (calibration, re-calibrating on a regular basis if necessary).

The goal here is not to make an extensive list of measuring devices, but to recall several useful concepts to guarantee good measurement results. Indeed, using digital devices equipped with modern automatic setting adjustment functionalities (it is no longer necessary to select a caliber) can make one lose sight of some elementary principles that have, however, been around since the invention of the first devices with pointers. Indeed, it is important to ensure that the caliber used to perform a measurement be the smallest

---

28 Although this technology is considered to be outdated, it still retains a good educational value for the understanding of measurement problems (manual calibre selection, reading uncertainty).

possible (without of course damaging the device) in order to maximize accuracy. This rule holds for a pointer device for which maximum deflection has to be obtained. And in the case of a digital device, the entry stage using an analog-to-digital converter (ADC), it is best for the measurement to be greater than the quantization step size (or *quantum*) in order to reduce its impact on the global measurement uncertainty.

Measurements preformed by analog devices are becoming increasingly rare (although displays of this type still exist). However, they also emphasize signal nature and therefore help select a suitable measuring device. This last point is particularly important because this family of analog measuring devices includes the following[29]:

– moving coil instruments (symbolized by a U-shaped (horseshoe) magnet and a coil on a moving frame);

– moving iron instruments (symbolized by an electromagnet with a moving iron core);

– electrodynamic instruments (using a coil on a moving frame and an electromagnet instead of the U-shaped magnet used by moving coil devices previously mentioned).

The symbols of these different instrument types are displayed in Figure 1.33. The operating principles of these devices, that are extensively described in [INS 15, EPS 15], of course have significant impact on measurable quantities. Indeed, moving coil instruments can only measure continuous quantities (and thus only mean values in general). In these conditions, they cannot be directly used to perform measurements under alternative current. In this case, a rectifier has to be used upstream of the conventional

---

29 This is not a complete list of all devices in this family.

measuring device[30]. However, it should be noted that the mean value of the rectified quantity is not generally considered relevant information. The root mean square (RMS) of the considered quantity is much more interesting. Unfortunately, the transition from one to the other is not unique and therefore it not possible to formulate a relationship between them unless a hypothesis is made on the waveform. Given that sinusoidal waveforms are the most common in industrial systems[31], the following relationship exists between the RMS and the mean of the rectified quantity:

$$X_{\mathrm{RMS}} = \frac{\pi}{2\sqrt{2}} \langle x_{rec} \rangle \tag{1.52}$$

where X_{RMS} is the sought variable (displayed by the device) whereas voltage x_{rec} is the rectified version of the measured quantity $x(t)$ (thus $x_{rec}(t) = |x(t)|$) of which the moving coil sensor measures the mean value $\langle x_{rec} \rangle$.

Figure 1.33. *Device symbols (moving coil on the left, moving iron in the center and electrodynamic on the right), source: [EPS 15]*

If the measured waveforms change, this type of device of course becomes inadequate (scaling should then be changed

---

30 A diode is then added below the symbol of the moving coil device of Figure 1.33.

31 This used to be the case in the past but it is less and less true for modern setups because of the prevalence of power electronic converters.

depending on whether the signal waveform becomes sinusoidal, square, triangular, etc.). In this case, moving iron devices are preferred because they supply true RMS values (they are therefore called TRMS devices for True RMS) independent of the measured waveform, provided that the frequencies involved do not exceed the working limits of the device (warning: the induced iron losses can damage this type of device for exceedingly high frequencies).

The concept of TRMS devices event exists for digital devices because not all of them could be termed TRMS. For example, out of the Fluke 27-II and the 28-II multimeters, that are handheld multimeters for industrial context, only the 28-II model is TRMS. Finally, it is important to notice that when desiring to measure a RMS value, measurements are less accurate for alternative quantities (AC – sinusoidal or not) than for continuous components (DC). This statement is corroborated by the characteristics of even the most precise devices such as the Keithley bench multimeter (see Figure 1.34).

1.5.2. *Operation mode selection and safety measures*

In order to carry out measurements in satisfying safety conditions, it is necessary to know the different categories and their associated usage conditions. Indeed, as can be seen in Figure 1.35, an inappropriate use can lead to the device exploding with the risk of severely wounding the user. This is particularly important to keep in mind when performing measurements on high power circuits that are obviously closest to the supply network. High power circuits are more dangerous to perform measurements on due to the fact that transient overvoltages are more important there than in downstream circuits that are, as for them, protected by surge suppressors. This section therefore provides a perfect transition with the following chapter that focuses on components ensuring the protection of electrical circuits in general and of electronic equipment in particular. Additionally it refers to the sixth chapter of Volume 1 [PAT 15a] which deals with the concept of insulation/isolation distance between the tracks of a printed circuit board along with the width of these tracks depending on the type of circulating current.

DC Voltage

ACCURACY (INPUT IMPEDANCE AUTO)

Range [1]	Resolution	Input Impedance [2]	Accuracy ±(ppm of reading + ppm of range)				
			24 Hour $T_{CAL} \pm 1°C$ [2]	90 Day $T_{CAL} \pm 5°C$	1 Year $T_{CAL} \pm 5°C$	2 Year $T_{CAL} \pm 5°C$	Temperature Coefficient [3]
100.00000 mV [4]	10 nV	>10 GΩ or 10 MΩ ±1 %	6 + 9	12 + 9	18 + 9	29 + 9	0.1 + 2.5
1.0000000 V [4]	100 nV	>10 GΩ or 10 MΩ ±1 %	4 + 1	9 + 2	15 + 2	26 + 2	0.1 + 0.5
10.000000 V [4]	1 μV	>10 GΩ or 10 MΩ ±1 %	2 + 0.7	9 + 1.2	14 + 1.2	22 + 1.2	0.1 + 0.05
100.00000 V [5]	10 μV	10 MΩ ±1 %	8 + 3	(18 + 5) [5]	(22 + 5) [5]	(30 + 5) [5]	(0.15 + 0.05) [5]
				35 + 5	40 + 5	45 + 5	2.0 + 0.5
1000.0000 V [4,6]	100 μV	10 MΩ ±1 %	8 + 3	(19 + 5) [5]	(23 + 5) [5]	(31 + 5) [5]	(0.15 + 0.05) [5]
				35 + 5	40 + 5	45 + 4	2.0 + 0.5

True RMS AC Voltage and AC Current

Function	Range [83]	Resolution	1-Year Accuracy: ±(% of reading + % of range) $T_{CAL} \pm 5°C$						
			3 Hz to 5 Hz	5 Hz to 10 Hz	10 Hz to 20 kHz	20 kHz to 50 kHz	50 kHz to 100 kHz	100 kHz to 300 kHz
Voltage [84]	100.0000 mV	0.1 μV	1.0 + 0.03	0.30 + 0.03	0.06 + 0.03	0.14 + 0.05	0.6 + 0.08	4.0 + 0.5
	1.000000 V	1 μV	1.0 + 0.03	0.30 + 0.03	0.06 + 0.03	0.14 + 0.05	0.6 + 0.08	4.0 + 0.5
	10.00000 V	10 μV	1.0 + 0.03	0.30 + 0.03	0.06 + 0.03	0.14 + 0.05	0.6 + 0.08	4.0 + 0.5
	100.0000 V	100 μV	1.0 + 0.03	0.30 + 0.03	0.06 + 0.03	0.14 + 0.05	0.6 + 0.08	4.0 + 0.5
	700.000 V	1 mV	1.0 + 0.03	0.30 + 0.03	0.06 + 0.03	0.14 + 0.05	0.6 + 0.08	4.0 + 0.5
Temperature Coefficient/°C (all ranges)	–	–	0.01 + 0.003	0.03 + 0.003	0.005 + 0.003	0.006 + 0.005	0.01 + 0.006	0.03 + 0.01

Figure 1.34. *Bench multimeter 7,5 digits Keithley DMM7510 and its AC and DC voltage measurements characteritics*

Figure 1.35. *Multimeter subjected to an overvoltage greater than its voltage withstand value (source: blog.formatis.pro/securite-mesures-electriques)*

In order to prevent failures that would be catastrophic for the equipment and dangerous for the user, a device capable of withstanding the overvoltages (more or less hostile) that can occur where connection is made to the power grid should always be used. There are four categories specified by the IEC61010-031 safety standards:

– category IV: power grid (primary source, overhead line and cable system, including distributing busbars and associated protection equipment against overcurrents);

– category III: industrial power grid (fixed installations concerning industrial distribution and circuits at the input of the electrical maintenance of buildings – technical column, elevator, etc.);

– category II: (handheld or domestic devices and equipment, mains power outlet);

– category I: special equipment or parts, follow-up to category II (telecommunications, electronic devices, etc.).

It is clear that category IV is the one for which the voltage withstand value has to be the highest because the power grid is the most hostile.

On the other hand, belonging to a category is not enough to characterize a measuring device: there exists different voltage ranges for each category and a device can belong to more than one category for different voltages. Table 1.4 displays tests results of overvoltages withstood by instruments of all different categories and with different nominal voltages (600 and 1, 000 V[32]). Actually, this is a IEC61010-1 standards extract that clearly defines test conditions for lower voltage equipment (50, 150 and 300 V). It is, however, clearly shown that tests are performed under precisely described conditions (amplitudes, number of impulse repetitions, internal source impedance/resistance) as is what is presented in the next chapter in section 2.1.3.1, in particular for protections against lightning bolts.

Category	Nominal working voltage (DC or AC RMS to ground)	Voltage peak impulse transient (20 repetitions)	Internal test source resistance
CAT I	600 V	2 500 V	30 Ω
CAT I	1 000 V	4 000 V	30 Ω
CAT II	600 V	4 000 V	12 Ω
CAT II	1 000 V	6 000 V	12 Ω
CAT III	600 V	6 000 V	2 Ω
CAT III	1 000 V	8 000 V	2 Ω
CAT IV	600 V	8 000 V	2 Ω

Table 1.4. *Equipment overvoltage test values (source: www.fluke.com/Fluke/ieen/Training/Safety/)*

REMARK 1.3.– The concept of category not only applies to devices but also to cables. It is therefore necessary to make sure that cables used are of good quality, undamaged and adapted to the device used. As a consequence, measuring devices should not be separated from their accessories. Or at the very least, since category (or categories) and voltage ranges are normally indicated via markings on cables as for

---

32 It should in fact be known that the IEC61010 standards only apply to devices adapted for voltages lower than 1, 000 V.

devices (see Figure 1.36), these cable markings should be respected.

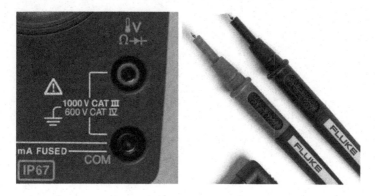

Figure 1.36. *Category/voltage range markings on measuring apparatus and wires (source: Fluke)*

2

Protections for Electronic Systems

2.1. Introduction and definitions

2.1.1. *Overview of malfunctions*

Generally speaking, protection components are designed to prevent electrical quantities from exceeding a certain threshold that would be dangerous for electrical or electronic device components. Exceeding this threshold could lead to catastrophic damage for the device. The major risks involved include explosion and fire hazard, with possible disastrous consequences on goods and people located nearby the failing device.

This actually leads straight back to the two fundamental quantities of electricity that are voltage and current. Evidently, damage to electrical or electronic equipment is generally a consequence of excessive energy dissipation over a given period of time. However, some processes can happen extremely fast. An example of this is the discharge phenomenon in an insulator which leads to the quasi-instantaneous complete or partial destruction of a component without qualifying as a case of excessive heating, unless considered on a very small scale.

Therefore, in this chapter, protections will be categorized into two families:

– protections against overvoltages and voltage surges;

– protections against overcurrents and short circuits.

The first case listed, i.e. overvoltages and surges, can be caused by:

– handling equipment on the power grid (part of the network offloading for instance);

– lightning (components are directly damaged or capacitive/inductive coupling occurs between the lightning bolt and the conductors);

– electrostatic discharges (particularly of human origin) commonly referred to as Electro Static Discharge (ESD).

More information on the latter two causes of overvoltage (although in this case they are more like voltage surges, i.e. short-lasting strong overvoltages) listed for electronic equipment can be found in Volume 4 [PAT 15d].

As for overcurrents and short circuits, these phenomenon both result in the circulation of excessive current. Before studying the possible causes of these two problems, it is necessary to define them further and to describe what distinguishes them:

– overcurrent: an overcurrent is an abnormality in an electrical (or electronic) circuit corresponding to a moderately excessive current;

– short circuit: a short circuit is an abnormality in an electrical (or electronic) circuit corresponding to a current greatly exceeding its nominal value.

Moreover, a short circuit happens very suddenly whereas an overcurrent is a slow phenomenon that can last between a few seconds and a few minutes. From these definitions, different scenarios that could eventually lead to these two situations can be suggested:

– an overcurrent could be the result of the moderate malfunction of an electrical (or electronic) circuit or of an over-consuming load (for example an electric motor) caused by the inappropriate use of a faultless device;

– a short circuit could be the result of a catastrophic component malfunction in an electrical (or electronic) circuit (for example, maintaining the closed state of a transistor when no control signal to do so is provided). Or it could be the result of a significantly altered function of the load powered by the electrical circuit (for example, mechanical block of an electric motor) due to the very inappropriate use of an initially faultless device (but that most certainly will not stay this way for very long without intervention!).

2.1.2. *Protections and standards*

Since they are protection bodies, components involved must prove their efficiency against performance standards. These standards are issued by various organizations throughout the world. So, test protocols are designed in order to determine whether or not components on offer by manufacturers meet the standard requirements imposed. Two organizations are particularly worth mentioning:

– the International Electrotechnical Commission (IEC), based in Geneva and that regroups many countries including (since 1906) the United Kingdom, the United States, France, Italy, Canada, Germany, Austria, Denmark, Sweden and Norway[1];

---

1 And ever since the very first assembly on the 26th and 27th of June 1906, Belgium, Holland, Japan and Spain send representatives.

– Underwriters Laboratories (UL) created in 1894 in Washington, USA.

The first organization listed is a gathering of national agencies that issue standards by common accord, concerning all matters related to electricity (and in particular related to protection bodies). As for the second organization mentioned, it is an American company specialized in safety certifications. Its seniority and expertise gave it *de facto* standards that are recognized and complied to by manufacturers. UL therefore plays a particular role that extends beyond the United States where another certification marking exists and is even mandatory for electrical devices: the Federal Communications Commission (FCC) label. Products complying with UL issued standards generally carry one of the two logos displayed on Figure 2.1. The logo on the left commonly applies to entire devices whereas the one on the right corresponds to single components.

Figure 2.1. *UL logos (for devices on the left and for components on the right)*

As for the IEC, it does not have a transnational logo to indicate standardized conformity of equipments. Additionally, countries are only members of the Commission through their own respective standardization agencies. Examples of such national standards agencies include:

– DIN & VDE for Germany (*Deutsches Institut für Normung and Verband der Elektrotechnik Elektronik, und Informationstechnik e.V.*);

– UTE for France (*Union Technique de l'Electricité*, merged with AFNOR (Agence Française de NORmalisation) since January 2014);

– ANSI for the United States (American National Standards Institute);

– British Standards Institute for the United Kingdom.

Figure 2.2. *A few conformity certification logos*

In reality, certification markings (Figure 2.2) are either linked to countries (for example, the FCC logo is linked to the United States as discussed previously[2]) or linked to groups of countries (CE marking for the European Union, EAC for Russia, Kazakhstan and Belarus). By placing these logos on their products, manufacturers are required to comply to the imposed standards and must be able to provide proof of compliance. For instance, in the case of the CE marking, the company must:

– obtain a certificate of conformity (from an approved organization);

---

2 This label is mandatory for all electronic equipment manufactured or sold in the United States.

– prepare technical datasheets for the product;

– sign a CE declaration of conformity.

The company must then be able to provide these documents at any time to the authorities upon request. In the case where the product has been imported, the importer is considered as responsible for product conformity. Indeed, the importer must make sure that the product manufacturer has carried out this procedure and he must gather the required documents. Furthermore, he must ensure that the competent authorities are put in relation with the manufacturer if necessary. Otherwise, an importer or a distributor can be a substitute for the manufacturer by affixing the CE marking themselves but for this the concerned party must first gather the required documents.

In order to allow electrical and/or electronic equipment manufacturers to ensure compliance of their products with standards, test equipment (overvoltage and ESD generators) have to be designed as advised by these standards organizations. This therefore guarantees a satisfying safety level for the user (see Figure 2.3). On the picture, the CE marking is clearly displayed (as the soldering station in question is commercialized in Europe), as does the VDE logo whose presence is expected for a product manufactured in Germany.

2.1.3. *Phenomena duration*

An overvoltage is naturally characterized by the voltage peak amplitude it reaches (this information should of course be completed by knowing the normal working voltage of the circuit that has to be protected). Similarly, an overcurrent or a short circuit is specified by the values reached by the current with or without the presence of a malfunction. These two families of defects have one thing in common: the temporal aspect (or time factor) of the malfunction whose duration can vary. In both cases, protection technologies implemented will be different depending on how much time is available to act. It is clear that a protection device must have a (much) slower response time than the time it takes for the malfunction to appear in order to be really efficient. Moreover, the response time should not be considered as intrinsic to the protection device only; the response time has an impact on the global design of the device that has to be protected and more particularly on the successful integration of the protection apparatus within the device. Indeed, a fast protection device used inappropriately can prove ineffective, or even harmful, to the (badly) protected device. In any case, having a good knowledge of the dynamics of malfunctions or of normal events capable of triggering untimely protective actions is required to correctly select protection bodies.

Figure 2.3. *An example of product marking (soldering station Weller)*

2.1.3.1. *Voltage surges, amplitudes and durations*

In the case of overvoltages, it is necessary to have extensive knowledge of the dynamics of typical voltage rises in order to know which surge suppressor is adequate. As a guideline, voltage spikes generated for ESD immunity tests last for a few dozen nanoseconds (less than 100 ns). However, overvoltages used to test equipment, connected to the power grid for example, have rise and fall times of the order of a dozen microseconds (making for a total duration of under 100 μs). Therefore, it is easy to notice that about three orders of magnitude separate the two timescales. As a consequence, this will have an impact on the adequate technologies to use for each of the two overvoltage categories (fast and slow).

As for ESDs, Figure 2.4 shows pictures of an ESD test gun. ESD guns are designed to generate pulses representative of those that could produce a human being charged with static electricity and and that could be discharged in a sensitive component through a finger (by direct contact or by simple proximity to the electric field generated). The used source model follows the guidelines of the IEC 61000-4-2 standard. Therefore, it behaves like a series RC circuit with a resistor $R = 330\Omega$ and a capacitor $C = 150 \,\text{pF}$ (in other words, with a time constant $\tau = RC$ of 49.5 ns). Tests for direct contacts are carried out under voltages of 2, 4, 6 and 8 kV whereas test for discharges to air (i.e. at a distance) are carried out under voltages of 2, 4, 8 and 15 kV.

Figure 2.4. *ESD immunity test gun TESEQ NSG 438*

Figure 2.5. *Voltage surge generator for photovoltaic panels EM Test VSS 500N6*

Specific generators are available to test equipment connected to the power grid and subjected to overvoltages directly applied to their connectors (e.g. mains power supply or telephone socket for telecom applications). These generators help ensure that this equipment meets certain standards. Generally speaking, there exist different generators depending on which application is targeted. For instance, Figure 2.5 shows a voltage surge generator dedicated to testing photovoltaic panels. This generator manufactured by the EM Test company under the reference VSS 500N6 is associated (according to the manufacturer's website[3]) with the following standards:

– EN 60255-5/IEC 60255-5: Electrical Relays – Part 5: Insulation coordination for measuring relays and protection equipment – Requirements and tests;

– IEC 61180-1: High-voltage test techniques for low voltage equipment – Part 1: Definitions, test and procedure requirements;

– IEC 62052-11: Electricity metering equipment (AC) – General requirements, tests and test conditions – Part 11: Metering equipment.

---

3 www.emtest.com.

These international standards are introduced here under their American heading because they are usually derived and translated in each country locally from the original publication. As a side note, they are also available for purchase on the websites of local standardization agencies (ANSI for American ones for example) and they are quite expensive (respectively, for the French versions, 97.90 ,208 and 118.80 , €, respectively, excluding taxes on the AFNOR[4] website at the time of writing – meaning on the 11th of April 2015).

However, it is worth mentioning that general standards exist for overvoltages to be applied on the connections of equipment that has to be tested (also called Equipment Under Test or EUT). They are further to the IEC 61000-4-2 standard on ESDs and are as follows:

– IEC 61000-4-4: Electromagnetic compatibility (EMC) – Part 4-4: Testing and measurement techniques – Electrical fast transient/burst immunity test;

– IEC 61000-4-5: Electromagnetic compatibility (EMC) – Part 4-5: Testing and measurement techniques – Surge immunity test.

Once again, these standards are introduced under their American heading as defined by ANSI. The headings clearly highlight the link that exists between these problematics and EMC (electromagnetic compatibility) field on which Volume 4 [PAT 15d] is focused. These two standards particularly define the voltage pulse waveforms to be applied in both cases but also the test bench organization required for certifying the equipment under test. Figure 2.6 displays, on its left, a surge pulse waveform defined according to the IEC 61000-4-5 standard, and on its right, a burst of pulses defined according to the IEC 61000-4-4 standard. The amplitude of applied overvoltages then varies depending on the device tested and

---

4 AFNOR = *Agence Française de NORmalisation*.

the assigned voltages in particular: for example, for a high voltage circuit breaker whose assigned voltage is 12 kV, the withstand voltage must be, according to IEC standards, of 28 kV at power frequency[5] and of 75 kV for a lightning impact.

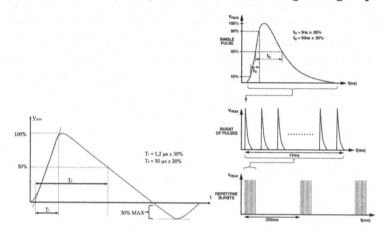

Figure 2.6. *Voltage surge waveform on the left (IEC 61000-4-5) and burst waveform on the right (IEC 61000-4-4)*

2.1.3.2. *Input surge current at switch-on of electrical and electronic devices*

Upon powering up an electrical or electronic device, the current absorbed from the power supply (battery or mains) is often greater than the nominal working current (steady state or even consumption during a power peak). This transient effect can have an impact on the selection process of the protection equipment in order to ensure that protective actions are not triggered every time the device is switched on (or frequently); this is particularly important for fuses. This large current absorption can also be termed "inrush current", or "input surge current" or again "switch-on surge". Actually, to counter this phenomenon, "shields" that apply specifically to the protection bodies of the device have to be implemented. Thanks to these "shields" the risk of unintentionally

---

5 Thus for a transient overvoltage due to a sudden consumption decrease (offlaoding) on the network for example.

triggering protection bodies is eliminated while avoiding making protections insensitive to very real malfunctions that could occur while the device is operating.

These protection techniques will also be discussed in the next section. They not only require the use of particular components but also of more or less complex setups combining several components. Existing solutions have both advantages and disadvantages which have to be weighed against one another in order to select the solution that is the most adequate for a given application based on technical and economical criteria. Nevertheless, a preliminary step consists in analyzing as precisely as possible the powered device to evaluate the amplitude and duration of the inrush current. Different types of loads are known to cause such inrush currents:

– transformers;

– electric machines;

– capacitive circuits (front capacitance rectifier for instance).

2.1.3.2.1. Transformers

Transformers were introduced and defined in Chapter 5 of Volume 1 [PAT 15a] under the assumption that their magnetic core had a linear $B(H)$ behavior. This is not exactly the case in practice: magnetic materials actually have a non-linear (i.e. saturated) and hysteretic behavior depending on frequency (losses by hysteresis – quasi-static – and Foucault currents). This study will consider the worst possible way a transformer can be used: operating it unloaded (i.e. zero current in the secondary). In this case the transformer is equivalent to a simple coil (the primary winding) with an iron core and as discussed previously, the modulus H of field $\mathbf{H}$ is proportional to current i (according to Ampere's law):

$$\int \mathbf{H}.dl = n.i$$

where n is the number of coils of the winding. Next, the voltage across the coil is equal to the derivative of flow B according to Faraday's law:

$$v = n\frac{d\varphi}{dt}$$

These two relationships always remain true (even in the case where the magnetic core has a non-linear and hysteretic behavior). If it is considered that the transformer is switched on at an arbitrary point in time, a differential equation has to be solved under the assumption that the initial current value is zero. Seeing as developing a non-linear equation is problematic, it is more pertinent to adopt a graphical approach here. The current waveform will therefore be analyzed using the plots displayed on Figure 2.7. It can be seen, qualitatively, that the current is almost in phase with induction $B(t)$ in the magnetic core. Furthermore, its waveform is deformed and exhibits spikes that are accentuated by the fact that there is a significant slope discontinuity between the linear operating region (having a high magnetic permeability $\mu = \mu_0.\mu_r$ and thus $\mu_r \gg 1$) and the saturation region in which the slope is close to the one in air (meaning $\frac{dB}{dH} \simeq \mu_0$). The case illustrated in Figure 2.7 is actually what happens in steady state in a transformer operating unloaded (i.e. with a secondary in open circuit). Indeed, it is generally considered that ohmic losses in the primary winding are negligible. Thus, the transformer operates under forced flow. Then, if a sinusoidal voltage is applied, since this voltage equals $n\frac{d\varphi}{dt}$ according to Faraday's law, flux $\varphi(t) = B(t).S$ is also sinusoidal. And since section area S of the magnetic circuit is constant[6], induction $B(t)$ is therefore also sinusoidal. This finding is corroborated by the plot of B against time shown on the top right of Figure 2.7.

---

6 The device studied here is a transformer and not a rotating machine: the magnetic circuit is therefore not deformable.

Next, in order to determine the temporal evolution of modulus $H(t)$ of field **H**, the value of $B(t)$ at each instant has to be carried forward on the *B(H)* characteristic curve of the magnetic circuit. Ultimately, this allows to reconstruct the temporal evolution of $i(t)$ which is proportional to $H(t)$ according to Ampere's law.

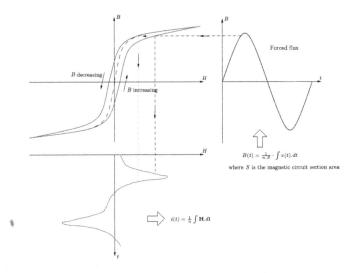

Figure 2.7. *Analysis of the magnetizing current in a real transformer operating unloaded. For a color version of the figure, see www.iste.co.uk/patin/power5.zip*

There is a trade-off when designing a transformer which generally leads to slightly saturate the core to avoid the need to oversize it and therefore to avoid impacting the total component weight. Therefore, in practice, real magnetizing current waveforms resemble quite closely the waveform plotted in Figure 2.7. It is however relatively easy to see that the situation can be significantly worsened in the transient regime directly following switch-on. Indeed, let us recall (after explicitly indicating integration bounds and setting $t = 0$ as the switch-on instant) that:

$$B(t) = \frac{1}{nS} \cdot \int_0^t v(\tau).d\tau \qquad [2.1]$$

and that voltage $v(t)$ can generally be written as:

$$v(t) = V_{\max}. \cos\left(\omega t + \phi_0\right) \qquad [2.2]$$

This leads to:

$$B(t) = \frac{1}{nS\omega} \cdot \left(\sin\left(\omega t + \phi_0\right) - \sin\left(\phi_0\right)\right) \qquad [2.3]$$

The case illustrated in Figure 2.7 corresponds to $\phi_0 = 0$ and therefore to a switch-on when $v(t)$ is at its maximum. In other cases, this leads to an induction, albeit sinusoidal, of amplitude $\frac{1}{nS\omega}$, but also containing a continuous component equal to $\frac{\sin(\phi_0)}{nS\omega}$. In the most detrimental case (i.e. with $\phi_0 = \pi/2$ and thus a switch-on with zero-voltage $v(t)$), this continuous induction component will be equal to the ripple amplitude. As for the choice of inducing a slight saturation at maximum induction in actual transformers, in this situation this would cause extreme saturation. Once the induction is defined, it is possible to look for the point corresponding to excitation H on the $B(H)$ characteristic curve in order to deduce the magnetizing current value. It is then obvious that this current becomes much greater than the current obtained in (symmetric) steady state. This situation is of course only transient: the induction eventually returns to a normal symmetric characteristic curve thanks to core losses. However, the return dynamics can last for a certain number of network current alternation (this happens much faster when the transformer supplies a load) and can apply significant stress on protection elements if they have not been designed to withstand this. In practice, this inrush current can last for a few seconds with an amplitude ten to fifteen times the nominal (primary) current amplitude of the transformer. It is also worth noting that this multiplier can be even greater in the case of toroidal transformers (up to sixty times).

2.1.3.2.2. Electric machines

Electric machines are also prone to significant switch-on surges. Generally speaking, when an electric machine is switched off, the impedance it presents to its power supply is weak and this leads to significant inrush current. The most obvious case is a direct current machine for which the armature behavior can be modeled as follows:

$$v = R.i + L\frac{di}{dt} + k.\Omega \qquad\qquad [2.4]$$

where:

– v is the voltage supply (in V);

– R is the armature winding resistance (in Ω);

– i is the current absorbed by this winding (in A);

– L is the winding inductance (in H);

– k is the electromechanical conversion constant[7] (in V/rad/s);

– Ω is the rotational speed of the machine (in rad/s).

When the machine is switched off, the model simply reduces to a simple series R, L circuit and if the nominal voltage U_N of

---

7 Which is only truly constant in the case of a permanent magnet machine. Otherwise, this coefficient is a function of the current circulating through an auxiliary winding called the excitation winding.

the machine is applied very suddenly, a current obeying the following function would be obtained:

$$i(t) = \frac{U_N}{R}\left(1 - e^{-t/\tau}\right) \text{ avec } \tau = L/R \qquad [2.5]$$

Actually, as soon as the current becomes null, a mechanical couple $c_m = k.i$ is produced and the machine starts up. Since the EMF $k.\Omega$ increases, the current then no longer increases and never reaches its asymptotic value $\frac{U_N}{R}$. However, it is clear that the inrush current duration heavily depends on how efficiently the machine speeds up and thus *on the driven load inertia*. This issue also obviously arises in synchronous machines which, even though they are alternative current machines, obey governing equations very similar to the ones obeyed by direct current machines[8].

Concerning asynchronous machines, even though their operating principle is quite different from that of synchronous and direct current machines, they still have a similar behavior regarding powering up processes in the case when power is supplied directly from the grid. Indeed, according to the layout described in Figure 2.8[9], a resistor of value R'_r/g appears where R'_r is supposed to be a constant resistance and g represents the machine slip ratio. This R'_r/g resistor is identified as a "motion resistor" and it is representative of the rotor movement with respect to the stator:

$$g \triangleq \frac{\omega_r}{\omega_s} = \frac{\omega_s - p\Omega}{\omega_s} \qquad [2.6]$$

---

8 Since these two types of machines are structurally symmetrical (simple role swap between the rotor and the stator).

9 Which is however obtained under the assumption of sinusoidal steady state power supply and therefore it is not rigorously applicable in a transient state case such as at switch-on.

where:

– ω_s is the angular frequency of stator quantities (voltages/currents) (in rad/s);

– ω_r is the angular frequency of rotor quantities (voltages/currents) (in rad/s);

– Ω is the mechanical rotational speed of the machine (in rad/s);

– p is the number of pole pairs of the machine.

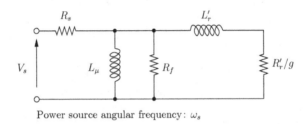

Power source angular frequency: ω_s

Figure 2.8. *Equivalent single-phase schematic layout of an asynchronous machine (short circuited rotor) in sinusoidal steady state*

In the schematic layout of Figure 2.8, all other elements are constant parameters (including the stator angular frequency ω_s required to calculate the reactances of the different inductances). Furthermore, the stator resistor R_s is often considered as negligible (*operation under forced flow* assumption). As a consequence, the circuit is simplified down to a *circuit with three branches in parallel* with a stator part comprising of a *magnetizing inductance* L_μ (as for a transformer) to which is associated a resistor R_f which is *representative of iron losses*. These two elements usually have a high impedance relative to the series combination of $L'_r, R'_r/g$ which is *representative of the rotor circuit* (short circuited winding coils or "squirrel-cage").

The slip ratio can easily be calculated at two characteristic operating points of the machine:

– when the machine is switched off, $g = 1$;

– once the synchronous speed is reached, $g = 0$.

As a result, during the start up phase, the motor goes from $g = 1$ to a near-zero slip ratio. This is because in practice, the asynchronous machine, even unloaded, never rotates at the synchronous speed but at an approaching value (even at nominal power for which the slip ratio is of the order of 5 to 10%. For example, a 1.5 kW machine with two pole pairs can reach a nominal value of 1 420 rpm when it is supplied with 50 Hz). These results lead to the conclusion that resistor R'_r/g can vary by a factor of 10 to 20 between switch-on and the stabilization of the machine speed at its nominal value. Of course, this finding has consequences on the impedance presented to the power grid. However, since this resistor is associated to an impedance $jL'_r.\omega_s$, its impact is weakened but it remains significant under a current of the order of six to seven times the nominal current.

The transient duration is once again specified by the nature of the mechanical load and in particular by its inertia. Several different approaches can be adopted to control this transient effect and limit the current and/or machine stress[10]:

– the star/triangle switch-on is the most common solution implemented in rustic applications that operate at constant speed;

– the progressive short circuit of a series of resistors at the stator;

– the progressive short circuit of a series of resistors at the rotor (in the case where a machine with a coiled winding is used);

---

10 Indeed, rotor copper losses are proportional to the slip ratio and conductors can get damaged if the switch-on phase lasts too long while being badly controlled.

– the use of an electronic starter (dimmer for fixed speed applications) or of a speed regulator (converter for variable speed applications).

This last solution proves that an efficient way to limit inrush currents consists in using a power electronic converter with an adequate control device (particularly for the response time).

2.1.3.2.3. Capacitive circuits

The situation that most frequently leads to significant inrush currents in the case of electronic equipment consists in connecting a decoupling capacitor that is initially charged to a (quasi) voltage source. In these conditions, the inrush current should theoretically be infinite, but actually it is limited by the loop impedance (inductance and resistance).

The inductor turns out to be a precious ally for the inrush current limiting battle seeing as it slows the rise of current. Unfortunately, it is also adverse to voltage because once it is associated to the decoupling capacitor, it will inevitably cause resonance (overvoltage) that could potentially be dangerous to the equipment downstream of the decoupling capacitor. Consequently, inductive limitation of the current is not a solution to prioritize. If it is nevertheless used, it must, at the very least, be extensively studied first and its implementation must incorporate dissipative elements capable of damping the resonance or a protection element protecting against overvoltages placed in parallel with the decoupling capacitor (and with the other devices connected to the latter).

Resistors are of course still present in capacitive circuits and must therefore be taken into account when calculating the intensity peaks at switch-on of a capacitive circuit. These resistors are those found in connections (contact resistors, wires, printed circuit board tracks) but also in components themselves: the ESR capacitor plays an important role in limiting the current and must be accounted for. Nonetheless, accounting for the ESR is not as simple as it seems because this capacitor parameter cannot be reduced to a unique constant value: it heavily depends on frequency (it tends to decrease as frequency increases for aluminium electrolytic capacitors[11]). This effect can impact results (with regards to experimental results) if this parameter is considered as constant for simulations of a voltage step test (which of course contains a wide range of frequency components). In any case, a simple model (of type series R,C or R,L,C) makes it possible to quickly obtain results in terms of determining the current peak amplitude and evaluating its duration. However this theoretical data will have to be confronted to experimental tests because it is most of the time not possible to fully comprehend and describe all setup parameters using a simple simulation (essentially due to insufficient capacitor data extracted from datasheets whereas inductors can be relatively well evaluated if the circuit geometry is known[12]).

2.2. Protection against overcurrents

2.2.1. Limiting inrush current

2.2.1.1. Resistors with negative temperature coefficient

NTC (for Negative Temperature Coefficient)-type thermistors are interesting components for current limiting in a circuit at switch-on (see Figure 2.9). Indeed, these components have significant resistance at room temperature and can, as a consequence, dissipate a significant amount of power when they are crossed by a given current. They are made of powdered metallic oxides bound by a plasticizing

---

11 At the very least in frequencies ranging from 50 Hz to 30 kHz, before the skin effect and the proximity effect lead to its increase for higher frequencies.

12 Parasitic inductances extracted by software such as Q3D Extractor (Ansys) for example.

agent. Due to heat dissipation, their temperature rises and their resistance simultaneously decreases.

Based on a careful component selection process, it is possible for the system to reach a thermal equilibrium such that the voltage across the thermistor remains weak when the device operates in steady state to deteriorate performance as little as possible. The upside of this solution resides in its simplicity that undoubtedly leads to a lessened price and a small-sized protective device.

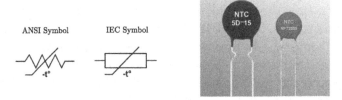

Figure 2.9. *NTC thermistor symbols and picture of two existing models (Hwalon Electronics Ltd)*

However, this operating principle does have several downsides that often turn out to be deal breakers:

– the device efficiency is necessarily deteriorated due to having an internal high temperature component and this solution will be difficult to implement in average or high power;

– the protection efficiency will be dependent on the ambient temperature surrounding the thermistor (ambient temperature sometimes highly variable in certain applications);

– after the device is switched-off, the thermistor needs time to cool down (at least one to two minutes) before it can be efficient again: an immediate switch-on right after switch-off makes this protection absolutely inefficient.

WARNING 2.1.– The NTC thermistor symbols shown in Figure 2.9 clearly highlight their negative temperature coefficient thanks the –t° indication. For a positive temperature coefficient (PTC) thermistor, the indication on their symbol is +t°. Nevertheless, in a number of schematics (such as the one shown on Figure 2.10, reproduced from an EPCOS guidance note), NTC and PTC thermistors are not always explicitly differentiated. In certain cases, the NTC or PTC term is indicated but in other cases, only the context of use makes it possible to determine the nature of the component.

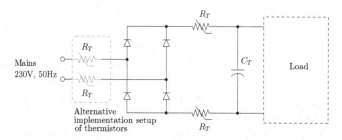

Figure 2.10. *Using NTC thermistors in a front capacitance rectifier*

The thermistor selection process is based on three key-parameters as shown by [EPC 13]:

– nominal resistance at 25°C called R_{25};

– maximal permissible current in continuous regime (I_{max} in DC or RMS in AC);

– capacitance C_T of the load that has to be powered (specified for a maximal working voltage).

This last aspect is particularly important because it is representative of the energy supplied to the capacitor during switch-on. As a result, thermistor selection is linked to a capacitance C_T for a given voltage and this capacitance has to eventually be adapted if the voltage used is different from the one indicated by the manufacturer for tests (see [DIG 15]).

For this purpose, the energy E simply has to be identified in both cases (with capacitance and voltage C_{T1} and V_1 on one hand and C_{T2} and V_2 on the other):

$$E = \frac{C_{T1}.V_1^2}{2} = \frac{C_{T2}.V_2^2}{2} \Rightarrow C_{T2} = C_{T1} \cdot \left(\frac{V_1}{V_2}\right)^2 \qquad [2.7]$$

The resistance at 25°C makes it possible to evaluate the current peak amplitude at switch-on since the thermistor senses the total voltage at this instant as the capacitor is entirely discharged. As a guideline, a typical application example is illustrated on Figure 2.10 where two NTCs are used at the input or at the output of a front capacitance rectifier connected to the 230 V and 50 Hz mains. Assuming that the placement selected is at the rectifier AC input, thermistors can be sized under the assumption that the nominal output current of the rectifier (which is assumed constant) is 1 A. In this case, the input current of the rectifier equals ± 1 A and has an RMS of 1 A. Furthermore, if it is considered that the highest voltage attainable is 230 V + 10 %, the RMS voltage prescribed for thermistor calculation will be equal to 253 V. From this result, the maximum capacitance that a thermistor can withstand can be re-evaluated according to equation [2.7]:

$$C_T = C_{T/230V} \times \left(\frac{253}{230}\right)^2 = 1,21 \times C_{T/230V} \qquad [2.8]$$

Therefore, relying on data shown on Figure 2.11 extracted from a NTC thermistor datasheet marketed by the TDK company, it is possible to notice that available references are able to withstand capacitances of $2,200\,\mu F$ under $230\,V$ in other words a bit more than $3,000\,\mu F$ under the voltage prescribed previously (230 V + 10 %). However, such capacitances are more than enough to filter out the rectifier output voltage when the current supplied to the load does not exceed 1 A (a maximum capacitance is indeed established at $1\,000\,\mu F$). Actually, thermistors suggested are substantially oversized because the maximum current of the smallest reference is equal to $8\,A$: it can therefore be hypothesized that the steady state resistance will be too high and is likely to deteriorate the rectifier efficiency. Another thermistor range introduced on Figure 2.12 turns out to be probably more adequate in particular since it has a thermistor that can withstand $900\,\mu F$ under $230\,V$ (i.e. a little over $1,000\,\mu F$ for the selected voltage) with a maximum current of $2.5\,A$.

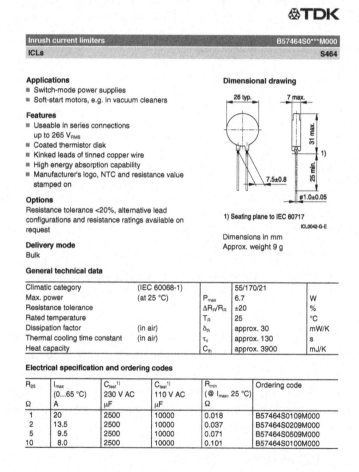

&TDK

Inrush current limiters	B57464S0***M000
ICLs	S464

Applications
■ Switch-mode power supplies
■ Soft-start motors, e.g. in vacuum cleaners

Features
■ Useable in series connections
  up to 265 V_{RMS}
■ Coated thermistor disk
■ Kinked leads of tinned copper wire
■ High energy absorption capability
■ Manufacturer's logo, NTC and resistance value
  stamped on

Options
Resistance tolerance <20%, alternative lead configurations and resistance ratings available on request

Delivery mode
Bulk

Dimensional drawing

26 typ.   7 max.
31 max.
7.5±0.8   25 min.
ø1.0±0.05
1) Seating plane to IEC 60717
ICL0042-G-E
Dimensions in mm
Approx. weight 9 g

General technical data

Climatic category	(IEC 60068-1)		55/170/21	
Max. power	(at 25 °C)	P_{max}	6.7	W
Resistance tolerance		$\Delta R_R/R_R$	±20	%
Rated temperature		T_R	25	°C
Dissipation factor	(in air)	δ_{th}	approx. 30	mW/K
Thermal cooling time constant	(in air)	τ_c	approx. 130	s
Heat capacity		C_{th}	approx. 3900	mJ/K

Electrical specification and ordering codes

R_{25}	I_{max} (0...65 °C)	C_{test}[1] 230 V AC	C_{test}[1] 110 V AC	R_{min} (@ I_{max}, 25 °C)	Ordering code
Ω	A	μF	μF	Ω	
1	20	2500	10000	0.018	B57464S0109M000
2	13.5	2500	10000	0.037	B57464S0209M000
5	9.5	2500	10000	0.071	B57464S0509M000
10	8.0	2500	10000	0.101	B57464S0100M000

Figure 2.11. *Datasheet extract of NTC thermistors of brand TDK*

REMARK 2.1.– The calculations carried out here lead to a unique thermistor. Actually, it is perfectly impossible to use two thermistors (as indicated on the schematic) sharing the capacitor charge energy at switch-on and being characterized by a steady maximum current even closer to the desired value (for example 1.5 A).

⚙TDK

Inrush current limiters					B57237S0***M0**
ICLs					S237

Electrical specification and ordering codes

R_{25}	I_{max} (0...25 °C)	C_{test}[1] 230 V AC	C_{test}[1] 110 V AC	R_{min} (@ I_{max}, 25 °C)	Ordering code
Ω	A	µF	µF	Ω	
1	9.0	700	2800	0.038	B57237S0109M0**
2.2	7.0	700	2800	0.064	B57237S0229M0**
2.5	6.5	700	2800	0.074	B57237S0259M0**
4.7	5.1	700	2800	0.120	B57237S0479M0**
5	5.0	700	2800	0.125	B57237S0509M0**
7	4.2	700	2800	0.172	B57237S0709M0**
10	3.7	700	2800	0.223	B57237S0100M0**
15	3.0	700	2800	0.346	B57237S0150M0**
22	2.8	700	2800	0.383	B57237S0220M0**
33	2.5	900	3600	0.507	B57237S0330M0**
60	2.0	400	1600	0.660	B57237S0600M0**

** = Delivery mode
   00 = Bulk
   51 = Reel packing

Figure 2.12. *Second datasheet extract of NTC thermistors of brand TDK*

It is clearly indicated in [EPC 13] that selecting an adequate thermistor has to be based on a limited ambient temperature (65°C in the previous example). Above this threshold, derating is required regarding the load for a given thermistor in order to guarantee that the component operates efficiently (see Figure 2.13).

2.2.1.2. *Other devices*

Other implementable solutions for inrush current limitation at switch-on are not as simple as NTC thermistors:

they are also neither as compact nor as cheap. However, they do make it possible to obtain better performances regarding:

– efficiency;

– setup heating;

– availability (even in the case the device is switched-on immediately after being switched-off);

– insensitivity to ambient temperature fluctuations.

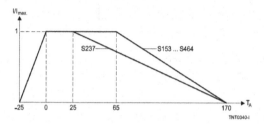

Figure 2.13. *Load derating (steady maximum current regarding ambient temperature as a function of the NTC thermistor family) (source EPCOS)*

An alternative to thermistors consists, among other things, in using a (relatively) temperature-independent resistor during switch-on, then shorting this resistor using a relay or an electronic switch (see Figure 2.14) once the voltage has risen. In these conditions, there is not only current limitation during switch-on but also absence of permanent resistor deteriorating the efficiency in steady state operation. *A priori* this solution is adapted to any sort of load (transformer, machine, capacitive circuit).

Nevertheless, a dissipative element is still in use with this solution while non-dissipative solutions exist for transformers and electric machines. For the latter machines, it is possible to use electronic starters based on dimmers or speed changers. Concerning transformers, it is also possible to use electronic setups called transformer switching relays (see Figure 2.15) whose basic principle is to synchronize the power-up with the supplied voltage value. In practice, this task is made more complex by the fact that the remanent field in the transformer magnetic circuit has significant impact on the inrush current at power-up.

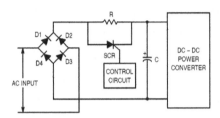

Figure 2.14. *Current limiting setup with a shortable resistance (source: [MOT 95])*

Figure 2.15. *Soft starter for motors (Allen-Bradley SMC-50) on the left and transformer switching relay (TSRLR of the FSM AG company) on the right*

In the case of capacitors, dissipative solutions are generally favored but, among these solutions, it is possible to find controlled solutions based on semi-conductors similar to those displayed on Figure 2.16 that are extracted from application note [MOT 95]. In both cases, the current flowing through the capacitor is limited at switch-on by the Drain-Source resistor R_{DS} of the transistor that operates linearly at switch-on and then tends to a very weak value R_{DSon} (thus in saturation) in steady state. The advantages of the two solutions are clearly identified. In the first case, the upside is that the equivalent series resistance (ESR) seen by the load is restricted to only the capacitor (contrary to solution no. 2 in which resistor R_{DSon} of the transistor is added to the capacitance in the ESR). And the second solution offers the advantage of preserving ground continuity between the input and the output of the device: there is therefore no shorting risk for the transistor in the event that the negative input terminal is connected, as is the negative load terminal, to the device ground (which is not the case in the first solution).

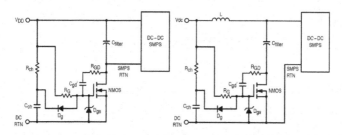

Figure 2.16. *MOSFET-based current limiting circuits for capacitors*

2.2.2. *Fuses*

Fuses are components dedicated to protection against overcurrents and short-circuits. They consist of a conducting element that is sized to resist a given maximum temperature rise beyond which it melts and interrupts the current. Conductor melting can be analyzed (as discussed in Chapter 6 of Volume 1 [PAT 15a]) using the Onderdonk equation. However the latter relies on the assumption that it operates adiabatically (i.e. with no heat exchange between the conductor and its surroundings): in this situation, no matter the current, the conductor always ends up melting after some time according to this model. In practice, this is of course not the case because of heat evacuation by conduction (through the fuse terminals towards connecting wires), by convection (through air) and by radiation. As a result, fuses are always characterized by a maximum working current for which the melting time tends asymptotically to infinity. Therefore, understandably, the key characteristic of fuses that allows to adequately select them in the first place is their response time as a function of the current that flows through them. Figure 2.17 shows an example of fuse specifications extracted from a Ferraz Shawmut datasheet.

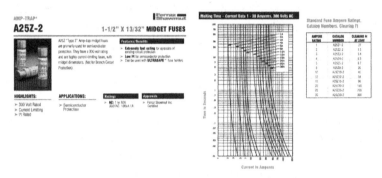

Figure 2.17. *Datasheet extract of a fuse for semi-conductor (Ferraz Shawmut)*

Reading through these specifications, it can be noted that for example a 1 A fuse will take between 90 and 10 s to melt under 2 A whereas it will take between 40 and 50 ms to melt under a current of 3 A.

Even though it is its main feature, the fuse must not only melt after being exposed to a certain current for a given time, it also has to be able to interrupt a circuit through which flows a very strong current (short-circuit) under a given voltage (which can be direct or alternative) and maintain this circuit interruption efficient after melting. Moreover, it is preferable that the fuse box remains intact after interruption. These many constraints are what force to conduct the adequate fuse selection process only after having carefully reviewed documentations. On the datasheet of the fuse range introduced on Figure 2.17, it is possible to see that, if current calibers range from 1 to 30 A, several specifications are common to all fuses, namely:

– the nominal working voltage (in this case 300 V AC);

– an interrupt rating (I.R.) of 100 kA;

– the $I^2.t$ rating that characterises the temperature rise that the fuse can withstand ($R.I^2.t$ being homogeneous for a dissipated energy).

It is also worth mentioning that the complete melting of a fuse is divided into two steps:

– a pre-arc phase during which the current increases and leads the fusible element to fuse;

– an arc phase during which the current can still increase (but has to end up decreasing until switch-off) and during which the current circulates in the air which is ionised because of the energy dissipated.

This electric arc is a necessary phenomenon to evacuate the energy stored in the circuit in magnetic form ($W_{mag} = 1/2L_{cc}.I^2$ where L_{cc} is the parasitic inductance of the shorted branch). In practice, choosing an adequate fuse entails correctly determining the short-circuit current of the device in question. The purpose of a fuse, regarding an electrical installation, is to prevent wire deterioration (and more precisely their insulators or even the mounting supports of the sets of bars in switchgear cabinets).

The most common fuse families are:

– gG: for general purpose;

– aM: motor protection fuse (those are time-delay fuses[13] supporting the large switch-on currents of motors – hence their name – as well as that of transformers[14]);

– uR: ultra-rapid fuse (those fuses are specifically designed for protecting semi-conductors).

It is worth mentioning that semi-conductor protection is also (as for conductor protection) conditional upon a trigger threshold that should prevent their temperature from rising too high. For example, diodes are characterized, as are fuses, by a $I^2.t$ ratio which must be greater (20 % greater according to [SOC 04]) than that of the fuse selected for their protection.

A great deal of useful information for fuse selection to ensure the protection of both electrical installations and electronic equipment, can be found in documents such as [SOC 04, FER 06, SCH 11]. This topic is indeed fairly complex due to the fact that, as for thermistors, ambient temperature plays a key role in fuse operation. Additionally, in applications specific to power electronics, fuses can be subjected to current pulses that can make them age prematurely even if the current RMS does not exceed component limits (see Figure 2.18).

Furthermore, operating under direct current is also more burdensome for fuses than operating under alternative current: indeed, interrupting an electric arc is more difficult

---

13 The time-delay makes this component too slow to provide efficient protection against moderately strong overloading. Therefore, thermal relays are used to ensure this type of protection for the power supply of industrial motors (since fuses are generally supported by the disconnectors in industrial systems).

14 aM-type fuses must be placed at the transformer primary while gG-type transformers must be placed at the secondary to ensure protection of the downstream circuit.

because current is not naturally stopped. All these aspects concerning applications specific to power electronics can lead to derating fuses and having to oversize them with regards to a more typical application on the 50 Hz AC power grid.

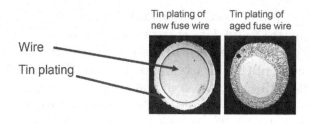

Figure 2.18. *Fuse ageing due to current pulses (source: [SCH 11])*

As for fuse packaging, there exist different fuse formats:

– the "cartridge" format with a glass or ceramic body. The meltable fuse thread is surrounded by air or by another insulating material which is also intended to absorb the heat dissipated by the melting thread and by the arc (generally silica powder);

– the "through-hole" format: this component is similar to the one above except it is destined to be soldered directly on a printed circuit board (cartridges, as for them, are placed in a fuse holder);

– the "surface-mounted" format (SMD): this is an alternative soldered on a printed circuit board more compact than the previous one. It only applies to small caliber fuses.

Cartridge fuses offer the advantage of allowing the user to replace them. Soldered versions (through-hole or SMD components) do not allow this. Therefore the manufacturer has to select fuse formats carefully depending on which type of user the device is destined for. In some cases, it could be a good idea to force the user to send the malfunctioning device back for servicing, instead of allowing the user to replace it with a new fuse that is very likely to suffer the same fate if the malfunction that melted the first fuse still has not be fixed.

WARNING 2.2.– An important piece of information concerning IEC and UL standards met by fuses is worth emphasizing. When a fuse is designed to withstand a given nominal current (1 A for example), it is capable of letting this current flow through 100 % of the time if it complies to the IEC standard. However, a fuse of the same caliber that meets the UL standard will only be able to let this current flow through 75 % of the time (see [SCH 11]).

2.2.3. Resettable fuses

Resettable or PTC fuses are positive temperature coefficient thermistors (see Figure 2.19) that play a complementary role to that of NTC thermistors studied previously. Indeed, the latter are responsible for limiting the inrush current at switch-on for properly functioning electric circuits whereas resettable fuses play a protection role against short-circuits and overloading which are characteristic phenomena of malfunctioning circuits.

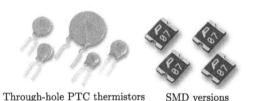

Through-hole PTC thermistors   SMD versions

Figure 2.19. *Pictures of resettable fuses: through-hole format on the right and SMD on the left (source: Farnell)*

Structurally, these components are comprised of an organic crystalline insulating matrix containing fine carbon particles (carbon black). In the crystalline state (when cold), carbon particles confer some electrical conductivity but if the current exceeds a certain threshold, the temperature rise caused modifies the matrix structure which goes from a crystalline state to an amorphous state. In this amorphous configuration, electrical continuity between carbon particles is interrupted and the component becomes an insulator. The component then only recovers its conductive properties once it has cooled down. This therefore is a current limiting component, and this is the objective sought for resettable fuses.

PTC thermistors are sometimes called resettable fuses but they are also known under many other trade names:

– *Polyfuse*: the trade name of components manufactured by the Littelfuse company;

– *Polyswitch*: the name belongs to the TE Connectivity company;

– other less frequently used trademarks exist. These include: *Semifuse* (ATC Semitec), *Fuzetec Resettable Fuse* (Fuzetec Technology) and *Multifuse* (Bourns, Inc.).

The application range of resettable fuses is very wide since calibers span over 20 mA to 100 A. These components are particularly useful in applications in which replacing typical fuses is difficult, if not impossible (spatial, nuclear applications) but also in "general public" applications such as switched mode power supplies as used in computers.

2.2.4. *Semi-conductor protections*

Fuses are relatively essential components in comparaison to semi-conductors thanks to their ability to interrupt circuits. Nevertheless, semi-conductors also have protection capabilities against overloading that are particularly interesting:

– they are fast;

– they are resettable;

– they make it possible to fine tune the triggering intensity (linked to the measurement accuracy of the latter) and to eventually implement a time-delay (tunable via the control unit).

These devices, called "solid state circuit breakers", are especially used in high voltage applications and can make use of various components such as very high power GTO thyristors. They are also found in low voltage applications, especially in batteries for electric vehicles (or even in radio-controlled models) as illustrated by the images in Figure 2.20.

Figure 2.20. *Examples of solid state circuit breakers for batteries (sources: Perfect Switch LLC, OBR Control Systems)*

2.3. Protections against overvoltages

Devices intended for protection against overvoltages are called "surge protectors" or "surge suppressors", and the term "surge" can sometimes be replaced by the term "spike" if talking about protective devices specifically designed to address particularly brief surges.

Three main component categories are used to address the different overvoltage types met in practice:

– gas discharge tubes (GDT) that make use of a discharge (breakdown) in a gas in order to dissipate energy and limit the voltage;

– varistors that are based on metallic oxides that have a non-linear resistivity as a function of voltage;

– semi-conductors used in cascades in the manner of Zener diodes (Transil diodes).

These three technologies are introduced in the following sections. Following this, a "protection" technique sometimes used in low cost equipment, which is based on the implementation of a pseudo-spark gap directly made of printed circuit board tracks, is mentioned in section 2.3.4. It will be seen that in this case, the protection is prone to uncertainties that fail to guarantee a satisfying degree of protection.

2.3.1. *Gas discharge tubes*

As indicated by their names, gas discharge tubes (GDT) enclose in their housing a sealed cavity filled with gas (possibly air) separating two electrodes between which an electric arc will form when sufficient voltage is applied. Indeed, the gas (any gas) has a feature called dielectric strength corresponding to electric field strength above which the gas loses its insulating behavior. Indeed, the field can be intense enough to strip away electrons from atoms and accelerate them until they reach a critical speed which is sufficient to lead them to strip away other electrons from the atoms with which they collide. This is a chain reaction in which the number of free electrons increases exponentially. This phenomenon is known as the Townsend discharge and falls under plasma physics. As a side note, GDTs are not always used as protective components against voltage surges. Therefore the term "GDT surge suppressor" is used to resolve this terminology ambiguity. If interested in a more thorough study of GDT variants available on the market, the reader can refer to [BOU 08]. However it is worth describing certain aspect of GDTs here.

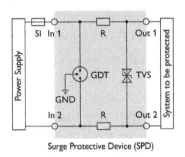

Surge Protective Device (SPD)

Figure 2.21. *Surge protective device schematic layout using (among other things) a three-terminal GDT*

First of all, GDTs are essentially intended to drain lightning energy between a line to be protected and earth (see Figure 2.21). Certain properties of these component are noteworthy to get to know their advantages and disadvantages better:

– on the one hand, they have extremely weak capacitance (typically 1 pF maximum, this prevents high frequency signals from being disrupted) unlike other protective components;

– they are, on the other hand, rather slow (it is common to find GDTs with a response time of 100 ns, if not more. These therefore partially let through quick surges);

– once triggered, they remain conductive under a voltage significantly lower than the voltage necessary for their initial triggering (as for fluorescent tubes used for lighting);

– certain models (see Figure 2.22) have an opaque housing (through-hole or SMD), others have a see-through housing (made of glass): these components, being sensitive to light (light influences the trigger threshold voltage), must be, if applicable, kept out of the light;

– two-terminal GDT models exists as well as three-terminal (or even more) models to allow to protect several lines simultaneously.

Figure 2.22. *Examples of GDT housing (from left to right): opaque through-hole (EPCOS) and SMD (Bourns) housings, see-through through-hole housing (Wikipedia, Lightning arrestor)*

Different versions of the GDT symbol exist but the most common represents electrodes using triangles pointing towards one another and separated by an empty space (see Figure 2.23).

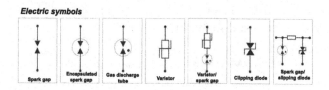

Figure 2.23. *Symbols of gas discharge tubes (and variants) and other surge suppressors*

As can be seen in the figure, the GDT symbol clearly shows the working principle of the component but the presence of a dot next to one of the electrodes does not suggest that the component is polarized since it is not the case. In fact, the dot indicates that the gas used is not air: this dot is absent in the case of a spark gap. Furthermore, the circle surrounding certain components symbolizes the sealed capsule in which their gas is contained therefore, understandably, the circle is absent in the case of the spark gap. The latter is simple a low cost version of the GDT and has limited efficiency. Spark gaps will be studied more extensively in section 2.3.4.

REMARK 2.2.– Different symbols exist but they are lesser known and used. One of them consists in replacing the two triangles representing the electrodes by the plates found in a capacitor symbol.

2.3.2. Varistors

Varistors (of which a symbol is shown in fourth position in Figure 2.23) are resistors whose value varies non-linearly as a function of voltage. They indeed show a high resistance under low voltage but, above a certain threshold, this resistance abruptly drops. These components are also called "metal oxide varistors" (MOV) because of their composition: they are made from a metallic oxide (zinc oxide for the most recent components) sintered to form a sort of ceramic (generally in the shape of a disc, as for NTC or PTC thermistors discussed previously).

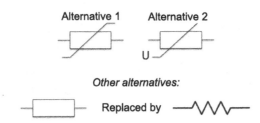

Figure 2.24. *Conventional varistor symbols*

The symbol introduced on Figure 2.23 is neither the only symbol nor the most commonly encountered. Different commonly used versions of the varistor symbol are displayed on Figure 2.24. The first alternative symbol shown is recognizable because it closely resembles the varistor symbol previously shown on Figure 2.23. Additionally, the second alternative symbol for varistors is similar to the thermistor symbol which is actually indicative of the fact that their resistance varies with "U" (i.e. voltage), while thermistor resistance varies with temperature (negatively for NTCs: −t°; positively for PTCs: +t°). Naturally, these symbols come in different shapes according to IEC standards (rectangle for resistors) and ANSI standards (zigzag symbol for resistors).

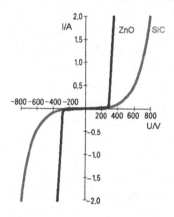

Figure 2.25. *Current/voltage characteristic curve of a zinc oxide based varistor (black curve) and a silicon carbide based varistor (gray curve). For a color version of the figure, see www.iste.co.uk/patin/power5.zip*

Figure 2.25 displays the current/voltage characteristic of a zinc oxide based varistor in comparaison to the silicon carbide based technology formerly used. It can be noticed that the break of slope at the triggering voltage (around 300 V) is much more abrupt with this new zinc oxide technology than with the former technology. Therefore this new technology makes it possible to obtain much more efficient components.

On the one hand, the response time of these components (of the order of 30 ns) makes them much more efficient than gas discharge tubes from the standpoint of the protection they provide. However, on the other hand, they have a much higher parasitic capacitance (100 pF to 1 nF for conventional

MOVs) which makes them incompatible with the protection of high frequency telecommunication lines[15]. Generally speaking, they are however well suited for protecting power supplies from mains with a very wide voltage range, ranging from tens of volts to kilovolts (and even beyond).

2.3.3. Transil diodes

Transil diodes, also called transient voltage suppression diodes (TVS diodes), are diodes similar to Zener diodes (in fact, they even sometimes are represented with the same symbol) making use of the same cascading phenomenon to limit voltage during overvoltages. TVS diodes therefore have a similar characteristic to modern varistors except they show a break of slope even more abrupt in the current/voltage plane which makes them even more efficient. This is especialy true since TVS diodes are extremely fast: their response time is indeed of the order of a few hundred picoseconds.

It should also be noted (see Figure 2.26) that this component is also available in unidirectional or bidirectional (two Zener diodes in series in opposite directions) versions. Protective devices that have been covered so far can now be listed in increasing order of response time:

– transil diodes (TVS);

– varistors (MOV);

– gas discharge tubes (GDT).

---

15 In reality, nowadays miniature "low capacitance" MOVs exist (of the order of a few picofarads at best) that can be used in telecommunication applications (for example the MHS range manufactured by the Littelfuse company).

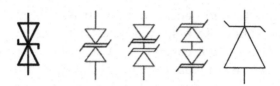

Figure 2.26. *Conventional TVS diode symbols*
(unidirectional on the right)

Following these findings, it is tempting to hypothesize that Transil diodes are the perfect protective devices. However, they are in general much less robust than varistors due to the fact that their ability to withstand surge energy is lessened at equal component size. Nevertheless, it must be said that malfunctioning Transil diodes generally behave like short-circuits. This guarantees that the input protected by this component is properly protected and that repair is possible by simply replacing the Transil diode.

From the standpoint of the parasitic behavior of TVS diodes, they also have a capacitance which is of the order of a few picofarads in the best case scenario and which can possibly exceed nanofarads for larger diodes[16]. As a result, Transil diodes show performances equivalent to that of MOVs specifically designed for high frequency circuits (while still being more fragile but providing greater protection than the downstream equipment!).

2.3.4. *Printed circuit boards*

One last protection device against overvoltages can be found in printed circuit boards of "low cost" items: they are called spark gaps. They are essentially like GDT suppressors

---

16 In reality, the capacitance value of a diode depends on the voltage applied: it can vary by several orders of magnitude when operating normally.

except they use non-encapsulated air as the gas to carry the discharge. They are mounted directly using pads (conductor is exposed – i.e. not covered by a resist) on a printed circuit board. Two examples of this technique are depicted in Figure 2.27. These images show an arc forming between two electrodes. The pads design makes it possible (in theory depending on the point shapes and the distance between electrodes) to set a threshold voltage beyond which a discharge will occur.

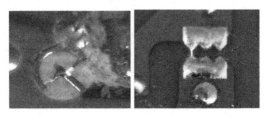

Figure 2.27. *Spark gaps on a printed circuit board (source: EEVBlog – eevblog.com)*

However, this technique is much less accurate than other techniques because the surrounding air is of variable pressure even for a domestic device and even more such for handheld equipment, or even equipment mounted in a vehicle. Indeed, the dielectric strength of air is significantly impacted by pressure and as a consequence it depends on the altitude at which the device is situated. Similarly, the moisture content plays a key role. As a result, all these factors (added to a possible contamination of the printed circuit board surface) make it impossible to guarantee a reliable reproducibility of the operation of this device. Spark gaps remains nevertheless an interesting component to implement when a solution as simple and cheap as possible is sought after (no costly component or expense associated to mounting the device on the board).

2.4. Explosion-proof equipment

2.4.1. *ATEX certification*

2.4.1.1. *Explosive atmospheres*

A major upside of power electronics is that they make it possible to replace mechanical switches, which are possible spark sources, by semi-conductors, which cannot generate sparks. This property turns out to be very useful in industrial environments where explosions are likely. Such areas are referred to as ATEX (from the French title of the 94/9/EC directive: *Appareils destinés à être utilisés en ATmosphères EXplosives* or, in English EXplosive ATmospheres). ATEX areas are indicated by the logo depicted on the left of Figure 2.28. Specific ATEX certified equipment (especially electrical equipment) is required for explosive atmospheres: they normally carry the logo depicted on the right of the same figure (although the background color can be different). The ATEX directive consists of two EU directives (94/9/EC or ATEX 137 for describing which equipment is allowed in ATEX areas, and 1999/92/EC or ATEX 100A for ensuring worker security).

Work areas are characterized in terms of gas or dust explosion risks and are classified into zones by employers:

– a type 0 zone corresponds to an area in which an explosive atmosphere comprising of a mixture of air and flammable substances in gas, vapor or mist form is continuously present or present for long periods of time or frequently;

– a type 1 zone corresponds to an area in which an explosive atmosphere comprising of a mixture of air and flammable substances in gas, vapor or mist form is likely to occur in normal operation;

– a type 2 zone corresponds to an area in which an explosive atmosphere comprising of a mixture of air and flammable substances in gas, vapor or mist form is not likely to occur in normal operation and if it occurs it will exist only for a short time;

– a type 20 zone corresponds to an area in which an explosive atmosphere in the form of a cloud of combustible dust is continuously present or present for long periods of time or frequently;

– a type 21 zone corresponds to an area in which an explosive atmosphere in the form of a cloud of combustible dust is likely to occur in normal operation;

– a type 22 zone corresponds to an area in which an explosive atmosphere in the form of a cloud of combustible dust is not likely to occur in normal operation and if it occurs it will exist only for a short time.

Consequently, the most dangerous zones are of type 0 or 20 whereas the least dangerous zones are of type 2 and 22.

2.4.1.2. *ATEX equipment*

ATEX certified equipment (especially electrical equipment) is defined with the help of a three part marking:

– the first part indicates the area location (I for mining applications, II for above ground industries);

– the second part indicates a category (1 for equipment allowed in zones 0 and 20 or less[17], 2 for equipment implementable in zones 1 and 21 or less and 3 for equipment implementable in zones 2 and 22);

– the third indicates the type of zone (G for gas-vapor-mist zones 0, 1, 2 and D for dust zones 20, 21, 22).

For example, equipment labelled II 2 G/D is allowed in above ground industries (in petrochemical plants for

---

17 Careful – the least dangerous zones are labelled with a higher number!

instance) in gas-vapor-mist zones 1 or 2 and in dust zones 21 or 22.

Figure 2.28. *Logos relative to the ATEX directive (area marking on the left and allowed equipment on the right)*

In the case of electrical equipment, a second marking completes this data. It indicates:

– the standards which the device complies to (E: for European norm CENELEC, EX: for IEC standards);

– the type of protection used: d = flameproof enclosure, e = increased safety, ib or ia = intrinsic safety;

– the gas groups (see Table 2.1): I, IIA, IIB, IIC in increasing order of hazard (for information: MIE = Minimum Ignition Energy; MESG: Maximum Experimental Safe Gap);

– the maximum surface temperature: T1 (450°C), T2 (300°C), T3 (200°C), T4 (135°C), T5 (100°C), T6 (85°C) from the less to the most constraining.

Groups	Typical gas	Gas groups	MIE (μJ)	MESG (mm)
Group I (mining)	Methane	I	300	1.14
Group II	Propane	IIA	240	0.92
(above ground	Ethylene	IIB	70	0.65
industries)	Acetylene	IIC	17	0.37
	Hydrogen		17	0.29

Table 2.1. *Classification table of explosive gases*

By way of example, the Fluke 28 II Ex multimeter (see Figure 2.29) holds the certification: IECEX ia IIC T4. It is therefore a device with intrinsic safety that can be used in the presence of the most hazardous gases and that guarantees a maximum surface temperature in normal operation of 135°C (which is actually rather impressive for a handheld multimeter!).

Figure 2.29. *Example of an ATEX certified electronic device (Fluke 28 II Ex)*

A few definitions are necessary to better understand the protective technologies implemented in an electric device in order to guarantee operational safety in an explosive atmosphere (especially the meaning of "intrinsic safety"). Actually, as specified by [INE 06], there are three ways to prevent an explosion:

– remove the explosive atmosphere;

– remove the ignition source;

– halt the spread of the fire.

In order to accomplish this, various approaches can be adopted. Concerning the first way to prevent an explosion (i.e. removing the explosive atmosphere), the enclosure of the apparatus (which is not tightly sealed[18]) can be held at an overpressure to prevent an explosive atmosphere from

---

18 It is difficult to obtain a perfectly tight sealing, especially regarding gases such as hydrogen. This is however not the case regarding dusts (as a side note, the Fluke 28 II multimeter is IP67 certified, which means it is water and dust tight).

entering the enclosure (this technique is also implemented to protect individuals against infectious agents using pressure suits). The electrical equipment can also be immersed in oil to isolate it from an explosive atmosphere. This commonplace solution is in fact frequently implemented for electrical equipment that is not situated in ATEX areas (such as power transformers for instance). Finally, an equivalent solution, easier to implement for handheld devices (such as the Fluke 28 II Ex), consists in encapsulating, sometimes partially, electrical circuits in a resin.

As for the concept of intrinsic safety, it is linked to removing the ignition source. In order to achieve this goal, the intrinsic safety device design guarantees that no spark, nor thermal effect, generated under conditions outlined by the standard, is capable of igniting an explosive atmosphere. This concept of intrinsic safety has to be distinguished from the concept of increased safety. For the latter, certain measures are taken in order to avoid, with an elevated safety coefficient, generating excessive temperatures and arcs or sparks inside and out of the apparatus in normal operation.

Finally, the last approach to prevent explosions is to prevent it from spreading (if an incipient explosion were to start inside a device). One of the first methods to achieve this consists in enclosing the device part susceptible of igniting an explosive atmosphere in an explosion-proof housing. This housing has to be designed to be resistant to explosion pressure while preventing the explosion from spreading. For this, the MESG (Maximum Experimental Safe Gap) concept introduced in Table 2.1 is important. The MESG describes the maximum gap size allowed in such an explosion-proof housing: if this gap size is exceeded, nothing stops the explosion from propagating. Another solution consists in filling the device (or at the very least the device part) susceptible of igniting the gas with a powdered material. The explosion propagating gaps then become either too small or too isolated to allow the explosion to develop further.

2.4.1.3. Temperature classification

In electronic devices, sparks (or electric arcs) are phenomena susceptible of setting off explosions. This issue can be bypassed by using electronic devices such as touch sensitive switches for human–machine interfaces and transistors or other components for power interfaces to replace mechanical switches and relays. However, heat remains a serious challenge in any electronic apparatus (even those making use of electronics).

Gases, vapors and dusts can be spontaneously ignited if temperature exceeds a given threshold. This threshold depends on the material to be ignited (see Table 2.2) and will have an impact on the minimum surface temperature allowed for a device exposed to a flammable product. This is the reason why different temperature classes have been defined (see Figure 2.3).

Product	Ignition temperature (in °C)
Carbon disulfide	102
Diethyl ether	170
Acetylene	305
Ethylene	425
Propane	470
Acetone	535
Hydrogen	560
Methane	595
Carbon oxide	605

Table 2.2. Ignition temperatures of various products (gases and dusts)

Therefore, it can be seen that in order to work in the presence of carbon disulfide, T5 or T6 class devices are required. For information, carbon disulfide is a dense, very volatile and very toxic liquid used as a solvent for many organic compounds. It is also used as a synthetic intermediary in the production of organic sulfur compounds.

Temperature classes	Maximum temperature (in °C)
T1	450
T2	300
T3	200
T4	135
T5	100
T6	85

Table 2.3. *Temperature classification*

2.4.2. *Protection index (IP) concept*

Although the concept of protection grade is not directly related to the ATEX directive, it is not totally neutral regarding dusts either. Indeed, the protection index characterizes apparatus tightness:

– the first digit in the rating specifies the protection against contact and foreign bodies (solid objects);

– the second digit specifies water-tightness.

Let us first identify the meaning of the numbers associated to the first index of the protection grade:

– 0: no special protection;

– 1: protected against foreign bodies with a diameter larger than 50 mm;

– 2: protected against foreign bodies with a diameter larger than 12.5 mm[19];

– 3: protected against foreign bodies with a diameter larger than 2.5 mm;

– 4: protected against foreign bodies with a diameter larger than 1 mm;

– 5: protected against dusts;

– 6: dustproof.

The numbers associated with the second index have the following meaning:

– 0: no special protection;

– 1: protected against water dripping vertically;

– 2: protected against water sprayed at an angle, up to 15° degrees from the vertical;

– 3: protected against spray water from any direction up to 60° degrees from the vertical;

– 4: protected against spray water from all directions;

– 5: protected against low pressure water jets from all directions (hose nozzle of 6.3 mm, distance 2.5 m from 3 m away, flow rate 12.5 l/min ±5 %);

– 6: protected against high pressure water jets from all directions (hose nozzle of 12.5 mm, distance 2.5 m from 3 m away, flow rate 100 l/min ±5 %);

– 7: protected against temporary immersion (up to 1 m deep);

---

19 It is assumed that it is not possible to access live parts for IP 2X certified electric equipment.

– 8 and over: protected against permanent immersion under pressure (beyond 1 m deep), submersible equipment + expanded capabilities (high pressure cleaning in particular).

By way of example, the Fluke 28 II Ex multimeter (as the non-EX version) is IP 67 certified: it is therefore dustproof and waterproof up to 1 m deep. This turns out to be useful information regarding the ATEX standard concerning dusts even though it is not useful concerning gases (tightness being harder to obtain because of this).

It should finally be noted that when one of the indices is not applicable, it is replaced by an "X". For instance, equipment placed in switchgear cabinets (for example: contactors, disconnectors, thermal relays, etc.) are generally IP 2X certified: this means that they do not allow their live parts to be touched with fingers. However, it is made clear that this type of equipment stops operating properly at the slightest water spray even sprayed vertically. This is nevertheless not a major drawback seeing as these components are anyway destined to operate in places shielded from external weather conditions and sprays of liquid.

3

Storage Systems – Principles, Technologies and Implementation

3.1. Introduction

The storage of electric energy is a difficult problem which can take on various forms depending on its applications and the ensuing constraints. If we take out "mechanical" energy storage (for instance, the kinetic energy of a flywheel, the potential energy of a pressurized gas or that of a water reservoir), the direct storages of electrical energy can be narrowed down to two large families of elements:

– the "kinetic energy" storage: coils;

– the "potential energy" storage: capacitors, super-capacitors and batteries[1].

---

1 In the case of batteries, we are in fact dealing with an electrochemical storage which is not exactly equivalent to capacitors and supercapacitors but which represents a very important part in electrical energy storage applications and which must therefore be analyzed.

The kinetic (electrical) energy storage consists of storing energy in magnetic form in a coil characterized by its inductance L thanks to circulation[2] of current i according to:

$$W_{\mathrm{mag}} = \frac{1}{2} L.i^2 \qquad [3.1]$$

We must, however, keep in mind that in order to work effectively, this type of storage needs to minimize the Joule losses, especially in the capacitors and must therefore use superconductors to do so. Implementing such solutions is difficult because of the extremely low temperatures that it requires. This difficulty makes this type of solutions industrially unusable and for the moment confines them to lab applications, where we can exploit their biggest advantage— their ability to provide or store significant amounts of energy in a very short time. In fact, energy storage is not, generally speaking, simply characterized by a quantity of energy stored per mass or volume unit, but also in relation to power: it is, in fact, useful to know the performance of a piece of technology in relation to another from this point of view, as well as how much time it will take for a certain amount of energy to be stored or extracted from a storage element without it being too taxing from a losses perspective (efficiency optimization) or even too stressful for the component (leading to a degradation of its functioning and to premature wear). In order to compare the different storage technologies in relation to these two criteria which are:

– the energy stored per mass unit (or energy density);

– the power that can be released by a storage element per mass unit (or power density).

We traditionally use a representation of functioning areas of the different technologies in a plane (power density, energy

---

2 Hence the "kinetic" storage quantifier.

density) named the *Ragone Chart* from David V. Ragone [RAG 68]. We can see an example in Figure 3.1.

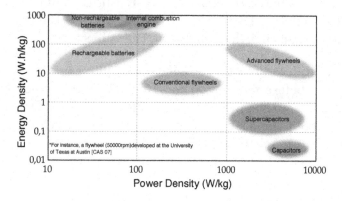

Figure 3.1. *Ragone Chart*

Let us note that the two axes are graduated in W/kg (axis X) and in W.h/kg (axis Y), both of them with logarithmic scales. It therefore becomes clear that the available technologies are complementary and that the electrical systems developer disposes of different tools for handling applications that need either large autonomies for a regular power demand, or for applications that need to provide intense power spikes in an episodic fashion.

Furthermore, this chapter will not deal with fuel cells which are not reversible equipments[3]: the notion of electrochemical battery will only be mentioned in the part dedicated to batteries, so as to make the distinction between reversible electrochemical devices and non reversible ones. Furthermore, the objective here is to present these components from the "user's" vantage point, or, more precisely, from the viewpoint of a developer of electronic power converters, who needs to associate the converter

---

3 hence the name of "cells".

technologies at their disposal to batteries (or supercapacitors), while also knowing the usage limits and constraints. Therefore, those interested in the subject of fuel cells can turn to [BOU 12], [REV 14] and [VIE 03] if they wish to deepen their knowledge on the subject.

First and foremost, we will now approach the case of capacitors before moving on to supercapacitors, which is the logical continuity towards more energy storage. One could inquire about the absence of inductances in this chapter (the classical ones, not the superinductances). The fact is that the subject has already been dealt with in a relatively comprehensive manner in Chapter 5 of Volume 1 [PAT 15a], which was dedicated to passive components: some information has also been provided on the capacitors, but this data is looked at in more depth (especially for polarized capacitors) and studied from the perspective of the transition towards supercapacitors. We will then look at batteries (rechargeable or not), still in the interest of logical (and historical) continuity-between the (relatively) non-reversible components – which have a poor energy storage capacity – and the reversible components, which are more and more widespread in modern electronic devices.

On the foundation of the physical characteristics of these components, we will end this chapter on their implementation within electronic power converters by inquiring about their monitoring. This proves to be a significant point in the good management of global systems, which we desire to be efficient and safe.

3.2. From capacitors to supercapacitors

3.2.1. *Capacitors*

Capacitors are components that ensure a (low) storage of electrostatic energy between two metallic frames (electrodes) subject to a difference in voltage noted V. Between these electrodes, a (dialetric) insulator – whose nature varies depending on the technology – contributes to the behavior of this component with its thickness e and permittivity $\varepsilon = \varepsilon_0.\varepsilon_r$ (where ε_0 is the vacuum permittivity and where ε_r is the relative permittivity of the insulating material that we have used). In the hypothesis of a geometry that can be assimilated to two large parallel planes in front of the thickness of the dielectric (insulator), the capacitance C expressed in Farads (F) of the capacitor is expressed thus:

$$C = \frac{\varepsilon.S}{e} \qquad [3.2]$$

where S is the surface of electrodes or of the metallic frames in front of it.

In order to obtain the highest capacitance possible, it is therefore necessary:

– to select the dielectric material whose relative permittivity is the highest possible;

– to increase surface S of the electrodes;

– to reduce the thickness e of the dielectric insulator.

We must, however, keep in mind that there are constraints which oppose (or could oppose) these three possibilities. Firstly, the relative permittivity of the dielectric insulator and its thickness are connected to the properties of the materials. Even if a certain material features a high permittivity, it must also feature electrical and mechanical properties that are compatible with the application considered: we must not only be capable of forming an insulating film with low thickness, but we must also ensure that its dielectric rigidity (or disruptive field) E_d is compatible with the voltage levels that we wish to apply to the terminals of the capacitors ($V_{\max} < E_d.e$). Consequently, thickness e and permittivity ε are parameters joint together by the properties of the dielectric. Finally, the surface S of the electrodes will have an impact on the bulk of the capacitor. It will therefore be, limited by the application, but we will see that for certain technologies, using porous materials will allow us to increase this parameter significantly.

Material	Relative permittivity ε_r	Dielectric rigidity (in kV/mm)
Polystyrene	$2.4 - 2.7$	19.7
Polypropylene (PP)	$2.2 - 2.36$	$30 - 40$
Polycarbonate (PC)	2.9	15
Polyester (PET)	$2.8 - 4.5$	300 (film)
Polyethylene (PEN, PPS)	3	$19 - 160$
Polytetrafluorothylene (PTFE)	2.1	$60 - 173$ (film)
FR4 (PCB)	$4.2 - 4.9$	20
Mica	$3 - 6$	118
Barium Titanate $BaTiO_3$ (ceramic)	$1200 - 10000$	2
Aluminum Al_2O_3	$9.3 - 11.5$	$10 - 35$
Tantalum oxyde (pentoxyde) Ta_2O_5	$25 - 50$	625

Table 3.1. *Characteristics of dielectric insulators powerly used in capacitors*

First of all, we can draw up a list of dielectric insulators powerly used in capacitors by noting not only their relative permittivity but also their dielectric rigidity. This data is grouped in Table 3.1. We can, however, note that despite there being a highly disruptive field (625 kV/mm), the tantalum capacitors are confined to low-voltage applications in the same way as the ceramic capacitors, whereas these two materials feature the highest relative permittivity (even very high for the barium titanate-based ceramic capacitors). In this specific case, we must keep in mind that there is a very large range of relative permittivity and that consequently there are very different variations in capacitors:

– the first class capacitors with poor capacitance (NP0 and C0G) whose value is precise enough and, above all, stable, depending on the voltage and the temperature;

– the second and third class capacitors (for instance, in the 2nd class, the ranges X5R, X7R or Z5U following the EIA denominations) whose capacitance may vary very strongly depending, on the one hand, on the voltage applied (see Chapter 5 of Volume 1 [PAT 15a]) and on the other hand on the temperature (the dependence on temperature being described by the name of the range – see Table 3.2).

Temperature range				Capacitance variation	
1st caract.	Min. temp.	2nd caract	Max. temp.	3rd caract.	Variation
Z	+10°C	2	+45°C	A	±1%
Y	−30°C	4	+65°C	B	±1.5%
X	−55°C	5	+85°C	C	±2.2%
		6	+105°C	D	±3.3%
		7	+125°C	E	±4.7%
		8	+150°C	F	±7.5%
		9	+200°C	P	±10%
				R	±15%
				S	±22%
				T	+22 Ã −33%
				U	+22 Ã −56%
				V	+22 Ã −82%

Table 3.2. *2nd class ceramic capacitors*

Generally speaking, polarized capacitors (which use aluminum or pentoxyde tantalum as dielectric insulator) feature high Volume capacities but are limited to (uniquely positive) voltages that are relatively low (less than 500 V generally speaking) whereas the capacitors made from plastic film feature low values but they can also support very high voltages (quite often a few kilovolts). In the case of ceramics, we obtain hybrid components that have a potentially high volumetric capacitance with low nominal voltages (such as the polarized capacitors) when we are dealing with non polarized capacitors (such as those made of plastic film). The ceramic capacitors with a high volumetric capacitance, in contrast, feature highly non-linear behaviors in relation to the voltage (a drastically lower nominal voltage than the value shown by the manufacturer which only proves effective for voltages close to zero).

From a physical viewpoint, the contribution of electrolytic capacitors is not only limited to the dielectric properties of the materials used (e.g. aluminum or tantalum pentoxyde). It also comes from the increasing of electrode surface S and the reduced thickness e of the dielectric. First of all, the thickness e is not obtained by extending the film (as in the case of capacitors made of plastic film) but via oxidation (so a chemical reaction) of an electrode. By using this method, we can therefore obtain an extremely fine layer: in the case of capacitors made of aluminum, the anode (i.e. an aluminum sheet) is covered with an aluminum layer (Al_2O_3) of the order of $1.5\,nm/V$. Thus, $63\,V$ a electrolytic aluminum capacitor has a dielectric thickness of $100\,nm$ Let us compare this to the several micrometers which we can (at best) obtain with plastic films. The principle is the same for the tantalum capacitors where we use an anode made of tantalum powder compacted through a sintering process, and which we cover with a fine insulating layer of tantalum penoxyde (Ta_2O_5).

At the surface level of the anode, we may assume that it is porous in the case of a tantalum capacitor because of its being made through sintering: it is therefore drastically larger than the cylinder shape obtained according to Figure 3.2 (on the right). In the case of the aluminum capacitor, the aluminum sheet used by the anode also has its surface extended via a process of chemical etching (*useless*) done at the time of its manufacturing (see Figure 3.2 – on the left). In both cases it is worth noting that this additional surface only has an effect due to the fact that the electrode is bathing in the electrolytic which acts as a cathode. In fact, the metal of the negative terminal of an electrolytical capacitor does not play this role and is limited to that of "current collector".

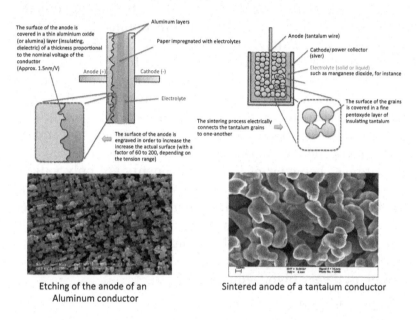

Etching of the anode of an
Aluminum conductor

Sintered anode of a tantalum conductor

Figure 3.2. *Description of aluminum and tantalum capacitors*

The electrolyte plays a major part in the well functioning of an electrolytic polarized capacitor:

– it acts as a cathode (and therefore as a conductor) in normal functioning mode;

– it ensures the possible regeneration (by oxidation) of the oxide layer at the level of the anode if this has just been damaged.

By contrast, it also contributes to the faults that the electrolytic capacitors present, to the extent that the ionic conduction is less effective than that of the electrons in a metal: it therefore introduces losses with the notion of equivalent series resistance (*Equivalent Series Resistance or ESR*) and can finish by degrading or evaporating and leaving the capacitor (with its faults in watertightness). The electrolytic capacitors are therefore subject to an aging process which will be more rapid as the functioning temperature will be higher[4]. To alleviate this problem, there are different types of electrolytes (solid electrolytes or frozen electrolytes) adapted to the needs of an application and in particular to the maximum functioning temperature (generally up to 150°C) in order to guarantee life expectancies of 1,000– 1,500 h with several tens of thousands of hours, depending on the components. It is, however, worth noting that the "capacitance" behavior of these components are only valid a few tens of kilohertz because, beyond that, their resisting behavior becomes dominant and when the frequency increases even more, we will see an inductive behavior characterized by a parameter marked ESL (namely, an equivalent series inductance). Generally speaking, the capacitors that can function at a very high frequency are:

– non-polarized;

– of a lower value;

– small in size[5].

In fact, all of the capacitors are characterized by a behavior resistant to intermediate frequencies and then by an inductive behavior at high frequencies. The impedance is

---

4 Even if the ESR of a capacitor has the tendency to reduce when the temperature increases.

5 For instance, the surface mounted devices (SMD) of type 0805, 0603 and 0402 as well as much shorter ones.

more or less difficult to model because it is not necessarily reduced to three parameters (C, R or ESR and L or ESL). In fact, the resistance of a capacitor may vary with the frequency: in this case, we tend to question the skin effect or the proximity effect between conductors, which involves an increase in resistance, depending on the frequency. However, the resistance of electrolytic capacitors reduces first and foremost with the frequency: the ESR of an aluminum capacitor is in fact higher at 50 Hz than at 1 or 10 kHz. This leads to proposing electric models that are more or less complex, as the ones in Figure 3.3.

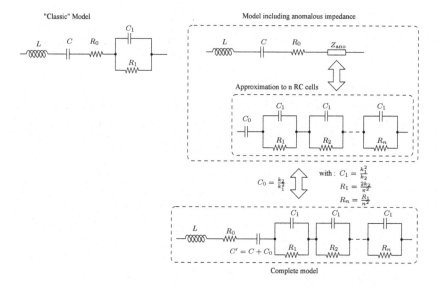

Figure 3.3. *Modelings ("classic" to the left, considering the diffusion to the right) of an electrolytic aluminum capacitor*

The equivalent left model is one of the most classic ones [GAS 05], but recent research has shown that the diffusive phenomena[6] to the implementation in the engraving of the anode[7] lead to introducing (see right hand model) an impedance referred to as "abnormal" (or at least a particular case) [COU 14] with the following expression:

$$Z_{\mathrm{ano}}(p) = \frac{k_1}{\sqrt{p}} \coth\left(\frac{k_2\sqrt{p}}{k_1}\right) \tag{3.3}$$

On this basis, we can propose an equivalent representation for which we replace this model of a non-integral order by a (serial) flow of parallel RC circuits. In fact, we have:

$$Z_{\mathrm{ano}}(p) = \frac{k_1^2}{k_2 p} + \frac{2k_1^2}{k_2} \sum_{k=1}^{\infty} \frac{1}{p + \frac{n^2\pi^2 k_1^2}{k_2^2}} = \frac{1}{C_0 p} + \sum_{k=1}^{\infty} \frac{R_k}{1 + R_k C_1 p} \tag{3.4}$$

We can therefore identify a serial (ideal) capacitor C_0 with an (infinite) set of parallel circuits $R_k C_1$, them also connected serially. The identification of the parameters leads us to the following results:

$$\begin{cases} C_0 = \frac{k_2}{k_1^2} \\ R_k = \frac{R_1}{n^2} \ \text{avec} \ R_1 = \frac{2K_2}{\pi^2} \\ C_1 = \frac{k_1^2}{k_2} \end{cases} \tag{3.5}$$

Obviously, for a practical exploitation, the infinite sum must remain limited to a certain number of cells. This choice depends on the range of frequencies for which we want a good adaptation of the model to the experimental impedance data. For example, we have obtained satisfying results with four or five cells for the Kemet PEG225 capacitors of $470\,\mu\mathrm{F}/63\,\mathrm{V}$ with

---

6 More precisely, we are dealing with a phenomenon of restricted diffusion where the ions stop at the bottom of the cavities made by the engraving because the oxide is supposed to be perfectly impermeable.

7 This research bears on the electrolytic aluminum capacitors.

the aim of monitoring the impedance via Kalman filtering within a static converter in order to evaluate their aging.

Let us note that this model does not account for self-charging phenomena in the sense that no resistance is placed in parallel with the main capacitance C_0. What is more, dielectric losses are considered with the flow RC cells. Another phenomenon known as the "battery effect" is not considered: this effect appears when we briefly short-circuit a charged capacitor. In this case, the voltage at its terminals drops to zero but, once the short circuit is suppressed, the voltage at its terminals is partially reestablished. This suggests an energy storage in chemical form, similar to that of a battery; this can, however, be modeled as an R-C serial circuit placed in parallel to the initial capacitor model with a high resistance, leading to a time constant such as the duration of the application of the short circuit that proves (very) insufficient for discharging this parasitic capacitance. All of this phenomena, to which we can add potential non-linearities (capacitance that varies with the voltage or the temperature, a hysteretic characteristic of certain dielectric materials, etc.) show us just how difficult a seemingly simple component can be to model.

3.2.2. *Physics elements in conductivity*

We can establish three types of classic conductivity in electric circuits:

– the conductivity in metals (circulation of free electrons);

– the conductivity in semi-conductors (circulations of electrons and holes);

– the ionic conductivity in the electrolytic capacitors, fuel cells and batteries.

We will not look here at the physics of semi-conductors, as the interested reader can learn about it here [MAT 01].

Therefore, we will only approach conductivity in ionic metals and solutions[8].

Conductivity in metals is addressed in the *Nernst–Einstein law* which refers to the migration of species in a crystalline solid. This law shows that the application of a force F will induce a gradient in the concentration c of species[9] but this concentration will homogenize by diffusion because of the Fick law. If we put this in the form of a problem1-D (following axis x), it translates as:

$$c_{\text{stat}}.v = D\frac{\partial c_{\text{stat}}}{\partial x} \qquad [3.6]$$

where D is the diffusion coefficient of the space considered in the crystal.

Considering that the force F is derived from a potential η ($F = -\frac{\partial \eta}{\partial x}$), the particles form a *Maxwell–Boltzmann distribution*[10]:

$$c_{\text{stat}} = c_0.\exp\left(-\frac{\eta}{kT}\right) \qquad [3.7]$$

where k is the Boltzmann constant ($k = 1.38 \times 10^{-23}$ J.K^{-1}) and T is the absolute temperature (in Kelvin). From here we can deduce that:

$$c_0.\exp\left(-\frac{\eta}{kT}\right).v = -\frac{D}{kT}\cdot\frac{\partial \eta}{\partial x}\cdot c_0.\exp\left(-\frac{\eta}{kT}\right) \qquad [3.8]$$

---

8 In fact, ionic conduction does not only exist in liquids. We frequently find frozen or pasty electrolytes in batteries or fuel cells. We can even find solid electrolytes (polymers, sintered powder in tantalum capacitors etc.).

9 In the case of metals, we are talking about free electrons.

10 This distribution relies on the hypothesis of an absence of interaction between the particles themselves. This applies to perfect gases and is a good approximation of real gases (in our case, we are talking about "electrons gases" in a metal) but it does not apply to liquids.

hence:

$$v = \frac{DF}{kT} \tag{3.9}$$

This equation is precisely the *Nernst–Einstein law* which, in our case, will apply to a force $F = F_{\text{élec}}$ connected to the electric field $E = -\frac{\partial V}{\partial x}$:

$$F_{\text{elec}} = q.E \tag{3.10}$$

But the flow j of charges q in the metal is connected to their concentration c_{stat} and to their speed v by the relation:

$$j = c_{\text{stat}}.q.v = \frac{c_{\text{stat}}qDF}{kT} \tag{3.11}$$

Remembering the Ohm law $j = \sigma.E$, we can establish the conductivity expression σ of the material:

$$\sigma = \frac{c_{\text{stat}}q^2 D}{kT} \tag{3.12}$$

We can therefore deduce from here that the conductivity decreases when the temperature increases. However, the resistivity ρ being the opposite of conductivity, we deduce from here that the metals (generally[11]) feature a positive temperature resistivity coefficient (CTP).

In the case of ionic solutions, the situation is different: the charges are carried by ions and not electrons. We are therefore dealing with bulkier elements, and therefore, significantly less mobile ones. In these conditions, we can expect a different electric behavior. From a qualitative perspective, the resistivity of a metal can be seen as the characteristic representative of collisions of free electrons with the crystal

---

11 Obviously we are only dealing with a model based on hypotheses that cannot constitute universal laws.

made by the atoms that are supposed to be fixed. In fact, these atoms vibrate more or less along with the temperature: the higher the temperature, the larger the amplitude of the movements of atoms around their balance positions. The interactions between the free electrons and the atoms[12] are then more probable and this can be translated into the macroscopic scale by an increase in resistivity. In the case of a liquid that contains ions, the mobility of the latter is higher as the temperature increases because of thermal agitation (Brownian motion): this can be compared to the behavior of parts placed in a vibrating carrier on an industrial assembly line. The ionic conduction also differs from the conduction in metals at a frequency level:

– the conduction in the metals is affected by the skin effect that reduces the effective section of a conductor. Consequently, the resistance increases when the frequency increases;

– in the case of ionic conduction, even if the phenomenon of skin effect cannot be eliminated, we can first notice a reduction in resistivity when we begin the analysis starting from low frequencies[13].

In order to analyze from a qualitative point of view this phenomenon of resistance that decreases when the frequency decreases, we should analyze the amplitude of the oscillating movements of the ions around their balance positions when an alternative electric field is applied. We will consider here that this amplitude is low enough for the interaction between the molecules to be insignificant. In the framework of a 1D

---

12 In the physics of materials, we speak of electrons/phonon interactions (see [GER 97]).

13 We are setting aside the case of DC current as this generally induces electrochemical reactions that modify the constitution of the electrolyte. The notion of impedance therefore partially loses its meaning in the sense that the doublet analyzed is not invariable.

problem, we can thus apply the fundamental dynamic principle to an ion with a mass M_i as follows:

$$F_i = M_i \frac{d^2x}{dt^2} \qquad [3.13]$$

and if we consider that the ion considered is subjected to a uniform electric field $E(t) = E_{max}.\cos(\omega t)$, the applied force F_i has the value:

$$F_i = z.q.E(t) \qquad [3.14]$$

where q is the elementary charge and z is the number of elementary charges carried by the ion considered. We can then assess the amplitude of the movement $x(t)$:

$$x(t) = \frac{z.q}{M_i} \iint E(t).dt^2 = -\frac{z.q.E_{max}}{M_i.\omega^2} \cos(\omega t) \qquad [3.15]$$

The amplitude of these oscillations is therefore inversely proportional to the square of the pulsation (and therefore the frequency).

Liquid	Molar weight M_m	Bulk density M_v (@ 20°C)	Size of the molecule	Intermolecular distance a
Water – H_2O	18 g/mol	998.3 kg/m³	1.515 Å	3.1 Å
Ethylene glycol – $C_2H_6O_2$	62.1 g/mol	1113.5 kg/m³	–	4.52 Å
g-Butyrolactone (GBL)	86.1 g/mol	$\simeq$ 1100 kg/m³	5.27 Å	5.1 Å
Dimethylformamide (DMF) – $(CH_3)_2NC(O)H$	73.1 g/mol	$\simeq$ 950 kg/m³	5.95 Å	5.04 Å
Dimethylacetamide (DMA) – $CH_3C(O)N(CH_3)_2$	87.1 g/mol	$\simeq$ 940 kg/m³	–	5.36 Å

Table 3.3. *Molar weights, molecular sizes, molecules and inter-molecular distances of several liquids*

Obviously, this behavior is purely theoretical as the amplitude of the movements is generally higher than the intermolecular distance: we can therefore expect a very different behavior because the interaction between molecules becomes inevitable. As a reference, we can see some characteristics of liquids in Table 3.3. The water is not the most widespread basic element in the constitution of electrolytes for electrolytic capacitors but it is however used for low-voltage components (100 V) and at a lowest rated temperature (85°C, 105°C respectively). The sizes of molecules (at least those that are available) are shown in Angström ($1\,\overset{\circ}{A} = 10^{-10}\,m = 0.1\,nm$) as well as the intermolecular distances which are deduced form the following formula:

$$a = \sqrt[3]{\frac{M_m}{M_v . \mathcal{N}_A}} \qquad\qquad [3.16]$$

where $\mathcal{N}_A = 6,022 \times 10^{23}$ is the number of elements (atoms or molecules) in a mole of matter. This is clearly an approximation which only gives one order of magnitude; but this shows that the ratio between the intermolecular distance and the size of molecules is close to 1 in all the cases where the comparison can be made. This result is clearly logical since a liquid is a thick medium compared to that of a gas. On these grounds, and by calculating the amplitude of the "free" oscillations of the ions whose expression is given to the equation [3.15], we see that the collisions will appear very soon, in fact as soon as the molecules are animated by an external electric fiend, regardless of the frequency used (usually 100 Hz Ã 100 kHz). Thus, we can expect that the electrolyte resistances have little or no dependency on the frequency. We notice, however, judging by the datasheets of electrolytic capacitors, that the increase in the frequency significantly reduces the ESR of capacitors. We can illustrate this result with the characteristics of the PEG 225 Kemet capacitor of $470\,\mu F/63\,V$ whose ESR is specified:

– 156 mΩ for 100 Hz and 20°C ;

– 52 mΩ for 100 kHz and 20°C.

We notice otherwise that this behavior is verified on all capacitors of the range (for all the capacities and all the rated voltages).

Figure 3.4. *Resistance measuring of a sodium chloride at ambient temperature (19 °C)*

In order to observe more specifically the behavior of an electrolyte (and not a conductor), we can easily put together an experiment so as to test the ionic conductivity depending on the frequency: it consists of measuring the resistance (between two electrodes) of a watery solution of sodium chloride – $Na^+ + Cl^-$ (which is table salt) with the help of a master RLC such as the Agilent U1733C. This is the easiest to produce ionic solution because it only uses common domestic products but it proves however perfect for our goal. We can see the realization of this experiment in Figure 3.4: the threads being rigid, we can easily keep them in place to take stable measurements. It is then possible to scan the frequencies with the button Freq of the master-RLC which allows us to take measurements at 100, 120, 1 k, 100 k and 100 kHz. The results of the measurements are grouped in the table in Figure 3.5 with the corresponding curve. They confirm a decrease in the resistance (and therefore an increase in the conductivity) at high frequency. In fact, in specialized literature (for instance [RAD 04]), the resistance of the electrolyte is considered independent from the frequency as we consider that the only interfaces with the electrodes induce impedances (of a parallel RC type) whose module reduces when the frequency increases[14]. In the RC cells associated to the two interfaces "electrolyte/electrodes" the resistance is qualified as *polarization resistance* whereas the capacitance is called *double layer capacitance* (which we normally find in the modeling of electrolytic capacitors, supercapacitors and even batteries).

Measuring data for different frequencies

Frequency	Resistance
100 Hz	31,6 Ω
120 Hz	29,2 Ω
1 kHz	17,3 Ω
10 kHz	14,7 Ω
100 kHz	14,1 Ω

impedance module *vs* frequency

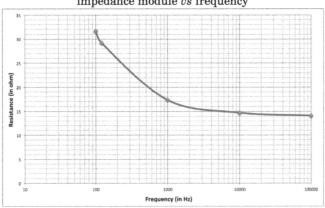

Figure 3.5. *Measurement results of the resistance of the sodium chloride solution depending on the frequency*

---

14 Because of the accumulation of ions at the surface.

For reference purposes, we can note that instruments dedicated to measuring the conductivity of electrolytes (for instance, those proposed by the Radiometer Analytical Society) use, to overcome the effects of the electrode/ electrolytic interface, a measure "with four terminals (or electrodes)" (just like the one for measuring the voltage at the shunts terminals of low values presented in Chapter 1).

3.2.3. *A few remarks about aging*

Capacitors are often considered fragile elements in an electronic equipment. This is particularly true for electrolytic capacitors where we consider that the liquid electrolyte tends to get away, particularly when the component is subjected to high temperatures (see Chapter 6 of Volume 2 [PAT 15b]). Different ranges of capacitors are however available depending on the functioning temperature: the most fragile capacitors (electrolytic aluminum) are designed to function at 85°C, other as 105°C whereas the most resilient can stand temperatures going as high as 150°C.

Generally speaking, capacitors can see their ESR increase along with their age. This is, however, not always the case and the loss in capacitance may be more sensitive (even for electrolytic aluminum capacitors). As a general rule, we may consider that a capacitor is in its last days when:

– its ESR has doubled (at the given functioning temperature[15]);

– its capacitance has dropped from 10 to 20 %.

This loss in capacitance is also a significant characteristic of non-polarized capacitors made of plastic film, as this is a self-healing process of the component that reduces the actual

---

15 Given that this parameter varies strongly between 25°C and the nominal functioning temperature.

surface of electrodes. We also notice this aging process in ceramic capacitors. On the contrary, for the latter, this loss is reversible and the capacitors can be regenerated by exposure to a high temperature.

From an operational viewpoint, aging is not always a continuous phenomenon and it can prove to be connected to functioning mishaps. All of the capacitors technologies are not the same in this respect. Indeed, certain technologies, which are *a priori* in good functioning order, can turn out to be sensitive and require an oversizing process. This is clearly the case of the tantalum capacitors compared to the aluminum capacitors from the point of view of voltage. When we choose a capacitor depending on its voltage caliber, we must know that an aluminum capacitor can be used at voltages as high as 90 % of this value whereas the safety margin required for tantalum capacitors can reach 50 %: this is therefore something to consider in a sizing process, just as we should consider the rates of the components (the aluminum capacitors being obviously a better choice in this regard). Let us remember that the voltage also plays a part in the behavior of class II or III ceramic capacitors with a large drop in the announced capacitance, as the voltage approaches the nominal voltage value of the component (see Chapter 5 of Volume 1 [PAT 15a]).

3.2.4. *Supercapacitors*

Supercapacitors are, as the name suggests, similar (at least in appearance) to capa and therefore characterized in the same way by a capacitance expressed in Farads. They are however different from the user's vantage point, by an extremely high volumetric capacitance and, therefore, nowadays we find supercapacitors whose capacitance reaches and even (considerably) surpasses 1 F. In contrast, their nominal voltage remains low (a few volts). Despite the fact that in a battery, the voltage is also low (at cellular level), the latter stores (regardless of its technology) higher energy densities thanks to an electrochemical energy conversion. Supercapacitors are however characterized by two major points:

– a longer life (in terms of the number of recharging cycles) compared to other batteries;

– a capacitance to either provide or store energy in very short periods of time.

This last point is perfectly illustrated in the Ragone chart presented in Figure 3.1 and it makes the supercapacitor an intermediary between the classic capacitors and the batteries in the range of electric energy storages in terms of power and energy. These characteristics (seen by the user) are the consequence of a design and functioning principle considerably different from those of classic capacitors and batteries.

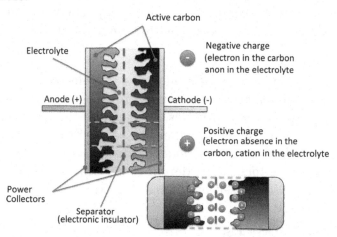

Figure 3.6. *Description of a supercapacitor*

In fact, the metallic electrodes of a supercapacitor are covered in active carbon (see Figure 3.6): it is, therefore, an extremely porous carbon structure which is impregnated with electrolytes. This electrolyte (essentially its ions) can circulate between the electrodes despite the presence of a separator which serves to electronically isolate the two electrodes. In the presence of an external electric field, the ions will migrate towards the electrodes and they will therefore spread out on the entire surface of the active carbon's pores[16] in order to compensate the external field and thus reach a state of balance. We must take note of the fact that the ions stay limited to the surface of the pores without interactivity with the carbon: there is no oxydo-reduction reaction taking place, in contrast to the case of batteries. This point is particularly important because it characterizes the difference between a supercapacitor and a battery (which stores the charges in the volume of the electrodes by "chemical transformation"). The storage being done in surface and not in volume, the energy density stored is lower than those of batteries. On the contrary, the charging and discharging can be done faster and no chemical reaction (or at least no "useful/researched" chemical reaction) contributes to the aging of the component: these are particularly the two strong points of the supercapacitors from the user's point of view.

Once the ions have migrated in the two electrodes, we can see a series association of two capacitors emerge, capacitors that are also charged in series. We refer to this as a double layer capacitance, which justifies one of the English names of capacitors: *Electrochemical Double Layer capacitor* (EDLC). In fact, supercapacitors bear different connected names, in fact to the denominations used by different makers (so here is non-exhaustive list):

– Ultracaps: EPCOS;

– Gold capacitors: Panasonic;

---

16 This is ion adsorbtion.

– Ultracapacitors: Maxwell   Technologies,   Skeleton Technologies;

– supercapacitors: Maxwell Technologies, Kemet;

– DL Cap (DL = *Double Layer*): Nippon Chemi-Con.

The global capacitance is therefore equal to the inverse of the sum of inverses of the two capacities: it is therefore smaller than the smallest of the two. The reason why the capacitance is finally high comes from the thickness of the dielectric that separates the opposing charges and from the surface that these charges come together on. The former is very low and is of the order of the size of the molecules making up the electrolyte whereas the second one is considerably increased because of the porosity of the material: in fact, the active carbon presents a surface higher than $500 \, m^2/g$.

The adsorption of anions (–) and cations (+) having been differentiated, the electrodes are not known in the same fashion: the component is therefore polarized (but for different reasons than those of the electrolytic capacitors). Finally, the maximum applicable voltage to a supercapacitor depends on the electrolyte used: we must at all costs avoid any chemical reaction and therefore we must not go past a certain threshold beyond which the solvent we have used might decompose.

The technologies of the supercapacitors depend, therefore, on two key factors:

– the nature of the active matter of the electrodes (active carbon, carbon derived from metallic carbides, etc.);

– the nature of the electrolyte used (aqueous liquid electrolytes, non organic/non aqueous electrolytes, ionic liquids).

The activated carbons are more or less porpous depending on their configuration and they are more or less costly but in order to obtain an optimum capacitance the cavities formed must be of the same size as the ions that will be received, namely of the order of 2 nm or less. The most commonly used material in power supercapacitors is the active carbon which is simple and not very costly to make. A carbon obtained from thick trees or from coconut shells for instance and where we have made plenty of cavities using a chemical/thermal procedure. The major problem of this material is a poor knowledge of the size of the pores made, which can have small diameters (micropores of a size < 2 nm) as well as more significant sizes (macropores between 2 and 50 nm).

Figure 3.7. *Supercapacitors using carbon derived from metallic carbides (Skeleton Technologies SkelCap)*

There are other, more costly materials, that allow us to obtain finer pores (between 0.6 and 1.1 nm for a specific surface of $1,000$ to $3,000 \, \text{m}^2/\text{g}$: this is carbon derived of metallic carbides or CDC (usually titanium carbide). The supercapacitors that use this technology are already commercialized by Skeleton Technologies (see Figure 3.7) in the industrial sectors researching high performances (the space and auto sections, for example). There are other solutions being researched in labs but, right now, they are not being implemented on an industrial scale. Amongst them, we can mention (see [SIM 13]):

– carbon nanohorns;

– carbon nanotube;

– graphene.

Regarding the electrolyte, the characteristics powerly being researched are:

– a good temperature performance;

– good ionic conductivity;

– reduced usage constraints (toxicity, flammability);

– good high voltage performance of the solvent.

This last point is particularly important since it plays an significant part in the energetic capacity of the supercapacitor. Indeed, we know that the energy stored W_{stock} can be expressed thus (just as a classic capacitor would):

$$W_{stock} = \frac{1}{2}CU^2 \qquad [3.17]$$

where U is the voltage applied to the terminals of the component. This voltage produces an electric field in the electrolyte and and that can be the place where several chemical reactions take place, when this field goes beyond a certain threshold. In the case of a liquid aqueous electrolyte, this voltage can not go beyond 1.2 V because beyond this, takes place an aqueous electrolysis which therefore produces hydrogen and oxygen within the component. Historically, the first supercapacitors used, however, aqueous electrolytes as they display a good ionic conductivity (which also means low losses). At the moment, organic electrolytes are still the most used: the solvent the most powerly used is the acetonitrile. It allows us to reach high voltages (of the order of $2.7\,V$) but has other inconveniences: it is toxic and flammable. This makes its usage problematic in the applications where security is a major criterion (particularly at high power because the volumes of the components become significant, making the equipments even more dangerous). We are powerly looking towards less toxic and less flammable solutions and a serious track os that of ionic liquids for which there are no solvents: these liquids are uniquely made of ions. This seems like an attractive solution but there are two major problems that stop them from being industrially carried through:

– viscous liquids that a high resistivity (especially at ambient temperature);

– these products are powerly very costly $(1,500\ \tilde{A}\ 2,000\,\$US/kg)$.

On the contrary, they would warrant the use of supercapacitors under voltages that can go beyond $3\,V$ and reach $5\,V$. In these conditions, following the equation [3.17], for a given capacitance C, going from 2.7 to $5\,V$ would lead to a multiplication of the stored energy by 3,4. For reference purposes, the Table 3.4 (made by [MER 13]) draws a comparison between the three large families of electrolytes that can be used in supercapacitors.

Electrolyte	Functioning voltage (V)	Conductivity© $(mS.cm^{-1})$	Functioning temperature (°C)
Aqueous electrolyte	1	> 400	from +5 to +80
Organic electrolyte	2.5	60	from −30 to +80
Ionic liquid	3–5	0.1–20	from −50 to +100

Table 3.4. *Characteristics of electrolyte families that can be used in supercapacitors*

3.3. From non-rechargeable to rechargeable batteries

3.3.1. *History*

The first cell is obviously the Voltaic cell which has been implemented by Volta in 1800[17]. It was made of a juxtaposition of metal discs (in this case, copper and zinc) separated by discs of felt imbued in salty water. Elementary devices of this kind were then stacked together (see Figure 3.8).

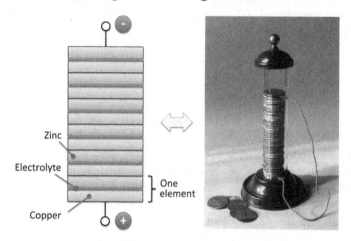

Figure 3.8. *Diagram and reproduction of the Voltaic cell (source: galvani-volta.e-monsite.com)*

In this cell, only the zinc and the water contribute to the chemical reaction that produces electrons for the electric circuit charged by the cell. The reaction at the level of the zinc is the following:

$$Zn \rightarrow Zn^{2+} + 2e^- \qquad [3.18]$$

---

17 In fact, cells may have existed since antiquity as pottery (see *the Bagdad cell*, dating back as far as 3rd Century BCE).

whereas in water, we have:

$$2H_2O + 2e^- \rightarrow 2OH^- + H_2 \qquad [3.19]$$

In fact, reactions depend on the nature of the electrolyte and in the majority of reference works that write about the Voltaic cell, the historic electrolyte is replaced with acid producing a more efficient cell. Generally speaking, the potential E of an electrode can be calculated in relation to the standard potential E^0 of the redox couple activated by the Nernst equation[18]:

$$E = E^0 + \left(\frac{RT}{nF}\right) \cdot \ln\left(\frac{a_{ox}^x}{a_{red}^y}\right) \qquad [3.20]$$

where R, T and F respectively are the constant of perfect gases (8.3144621 J.mol^{-1}.K^{-1}), the absolute temperature in Kelvins and the Faraday constant (96, 485 C.mol^{-1}). As for the parameter n, it represents the number of electrons transferred in the following semi-reaction:

$$x\,Ox + n\,e^- \rightleftharpoons y\,Red \qquad [3.21]$$

At 25°C, this equation is simplified, noting that $\frac{RT}{K}\ln 10 \simeq 0.059\,V$ and by assimilating the chemical activities a_{ox}^x and a_{red}^y of the oxidizer and of the reducer to their respective concentrations $[ox]^x$ and $[red]^y$:

$$E = E^0 - \frac{0.059}{n}\log\left(\frac{[red]^y}{[ox]^x}\right) = E^0 + \frac{0.059}{n}\log\left(\frac{[ox]^x}{[red]^y}\right) \qquad [3.22]$$

We must then classify the oxydo-reduction (or *redox*) couples depending on their standard potential. In fact, this is the basis of the calculation of the potential difference (or voltage) produced by a cell (or even an accumulator). We may see in Table 3.5 a list of redox couples with their standard

---

18 Which is not the same as the Nernst–Einstein law previously mentioned in the study of metal conductivity.

potentials E^0 at 25°C. Outside of the temperature, the measuring conditions have consequences on the actual concentration (1 mol/l) and on gas pressures (1 atm = 101.325 kPa). We will otherwise note the state in which we find the products: (s) for solids, (l) for liquids, (g) for gas, (aq) for liquid aqueous solutions[19]. Before we move any further, let us remember the definition of an oxidizer and that of a reducer:

– an *oxidizer* is a specimen capable of receiving either one or several electrons;

– a *reducer* is a specimen that may give away one or several electrons.

Oxidizer	Reducer	Semi-Equation	Standard potential E^0 (in V)
Li^+	$Li_{(s)}$	$Li^+ + e^- \leftrightharpoons Li_{(s)}$	−3.0401
$N_{2(g)} + 4H_2O$	$2NH_2OH_{(aq)} + 2OH^-$	$N_{2(g)} + 4H_2O + 2e^- \leftrightharpoons 2NH_2OH_{(aq)} + 2OH^-$	−3.04
K^+	$K_{(s)}$	$K^+ + e^- \leftrightharpoons K_{(s)}$	−2.931
Mg^{2+}	$Mg_{(s)}$	$Mg^{2+} + 2e^- \leftrightharpoons Mg_{(s)}$	−2.372
Mn^{2+}	$Mn_{(s)}$	$Mn^{2+} + 2e^- \leftrightharpoons Mn_{(s)}$	−1.185
H_2O	$H_{2(g)} + 2OH^-$	$2H_2O + 2e^- \leftrightharpoons H_{2(g)} + 2OH$	−0.8277
Zn^{2+}	$Zn_{(s)}$	$Zn^{2+} + 2e^- \leftrightharpoons Zn_{(s)}$	−0.7618
Fe^{2+}	$Fe_{(s)}$	$Fe^{2+} + 2e^- \leftrightharpoons Fe_{(s)}$	−0.44
Ni^{2+}	$Ni_{(s)}$	$Ni^{2+} + 2e^- \leftrightharpoons Ni_{(s)}$	−0.25
Pb^{2+}	$Pb_{(s)}$	$Pb^{2+} + 2e^- \leftrightharpoons Pb_{(s)}$	−0.13
H^+	H_2	$2H^+ + 2e^- \leftrightharpoons H_2$	0
Cu^{2+}	Cu^+	$Cu^{2+} + e^- \leftrightharpoons Cu^+$	+0.159
Cu^{2+}	$Cu_{(s)}$	$Cu^{2+} + 2e^- \leftrightharpoons Cu_{(s)}$	+0.340
$O_{2(g)} + 2H_2O$	$OH^-_{(aq)}$	$O_{2(g)} + 2H_2O + 4e^- \leftrightharpoons 4OH^-_{(aq)}$	+0.40
Cu^+	$Cu_{(s)}$	$Cu^+ + e^- \leftrightharpoons Cu_{(s)}$	+0.520
Ag^+	$Ag_{(s)}$	$Ag^+ + e^- \leftrightharpoons Ag_{(s)}$	+0.7996
Hg^{2+}	$Hg_{(l)}$	$Hg^{2+} + 2e^- \leftrightharpoons Hg_{(l)}$	+0.85
$Ag_2O_{(s)} + 2H^+$	$2Ag_{(s)} + H_2O$	$Ag_2O_{(s)} + 2H^+ + 2e^- \leftrightharpoons 2Ag_{(s)} + H_2O$	+1.17
$MnO_4^- + 4H^+$	$MnO_{2(s)} + 2H_2O$	$MnO_4^- + 4H^+ + 3e^- \leftrightharpoons MnO_{2(s)} + 2H_2O$	+1.70
$HMnO_4^- + 3H^+$	$MnO_{2(s)} + 2H_2O$	$HMnO_4^- + 3H^+ + 2e^- \leftrightharpoons MnO_{2(s)} + 2H_2O$	+2.09
$F_{2(g)} + 2H^+$	$2HF_{(aq)}$	$F_{2(g)} + 2H^+ + 2e^- \leftrightharpoons 2HF_{(aq)}$	+3.05

Table 3.5. Redox *couples and their standard potentials at 25 °C*

---

19 This is the default situation for all kinds of charges (both ions and electrodes).

On this basis, an oxydo-reduction reaction is thermodynamically possible if it satisfies the gamma rule, illustrated in Figure 3.9. We must however note that if a reaction is possible, it is not certain: it might potentially not take place or maybe take place extremely slowly (for instance, the oxidation of iron by the dioxygen is a very slow reaction).

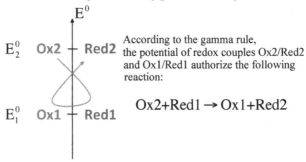

Figure 3.9. *Gamma rule*

3.3.2. *Daniell cell*

The Daniell cell is an improvement off the Voltaic cell which was invented by the British John Daniell in 1836. It exploits the two electrodes made of zinc and copper in order to produce a difference in potential of $1.1\,V$ (see Figure 3.10). For this, the zinc anode is submerged in a solution of zinc sulfate ($ZnSO_4/H_2O$) and the copper cathode is submerged in a copper sulfate solution ($CuSO_4/H_2O$).

Reactions at the level of the two semi-cells are thus the following:

$$Zn \rightarrow Zn^{2+} + 2e^-$$
$$Cu^{2+} + 2e^- \rightarrow Cu$$

[3.23]

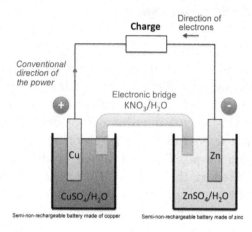

Figure 3.10. *Structure of the Daniell cell*

The electrons produced at the anode and received at the cathode circulate within the circuit charged by the cell. The circuit is otherwise closed between two solutions where electrodes are submerged with the help of an electrolytic bridge (a saline bridge) containing potassium nitrate (a solution containing the ions K^+ and NO_3^-). When power is provided by the cell, the ions NO_3^- and K^+ derive towards zing and copper sulfates solutions:

– for each ion produced Zn^{2+}, two ions NO_3^- are levied from the electrolytic bridge towards the zing sulfate solution;

– at the same time, an ion Cu^{2+} is transformed in a copper atom Cu on the cathode: two ions K^+ are then levied from the electrolytic bridge towards the zing sulfate solution;

REMARK 3.1.– In chemistry, the terms of the anode and cathode applied to a cell take an opposite way to that powerly applied in electronics. Indeed, the anode is the terminal whose potential is the lowest (negative) whereas the cathode is the terminal with the highest potential (positive).

3.3.3. *Power cell technologies*

3.3.3.1. *Saline cells*

The power cells are characterized by an absence of liquid. The closest technology of the Daniell cell still used nowadays is the Leclanché cell (1867), after its inventor Georges Leclanché. It is also called saline cell or dry cell insomuch as the electrolyte used (ammonium chloride – ions NH_4^+ and Cl^-) is present as gel. The two active products are the zinc, which is osidated at the level of the anode (which is the "–" terminal) and the manganese dioxide MnO_2 which is reduced at the level of the cathode (the terminal "+"). The reactions are the following:

$$Zn_{(s)} \rightarrow Zn^{2+}_{(aq)} + 2e^- \qquad [3.24]$$

and:

$$4MnO_{2(s)} + 2H^+ + 2e^- \rightarrow 2MnO(OH)_{(s)} \qquad [3.25]$$

The standard potentials of these two reactions are $-0.7618\,V$ and $+1.01\,V$ respectively. The voltage should therefore be $1.77\,V$ but in fact, the cell only delivers $1.5\,V$ because it only functions in standard conditions. Structurally, the positive electrode is made with a graphite cylinder (current collector) submerged in a manganese dioxide becomes impregnated with electrolytes and mixed with carbon powder in order to improve the electric conductivity between the electrolyte and the active matter (see Figure 3.11).

We note the presence of ions H^+ in order to ensure the reaction [3.25]: this is made possible by the electrolyte because the presence of ammonium ions NH_4^+ makes the solution acidic. The energy capacity of the cell being linked to the quantity of active matter in presence, it must be designed so that the limiting factor be of manganese dioxide and not zinc because zinc makes up the exterior wall of the cell (if we exclude the external insulating material that covers it). A complete consumption of the zinc would cause a parasitic reaction that produces dihydrogenate: different techniques allow us to fight this phenomenon (mixtures, protective organic elements). We note that these cells have been outdone by alkaline cells[20] more high performing.

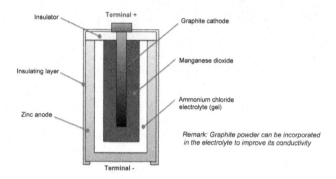

Figure 3.11. *Structure of the Leclenché cell*

3.3.3.2. *Alkaline cells*

Alkaline cells (or Mallory cells) have been known throughout the second World War. The active materials are always (as for saline cells) the zinc and the manganese dioxide but the structure of the cell and electrolyte are different (see Figure 3.12).

In fact, here we use *potassium* which contains K^+ ions and OH^- ions. Consequently, the electrolyte is a basic medium (pH $\simeq$ 13). A basic medium being qualified as alkaline, this electrolyte has given its name to this cell technology. At the level of electrodes, the reactions are as follows:

$$Zn_{(s)} + 2OH^-_{(aq)} \rightarrow ZnO_{(s)} + H_2O + 2e^- \qquad [3.26]$$

---

20 Which powerly dominate the market by representing, in France for instance, 75 % of cell sales.

and:

$$2MnO_{2(s)} + H_2O + 2e^- \rightarrow Mn_2O_{3(s)} + OH^-_{(aq)} \hspace{2cm} [3.27]$$

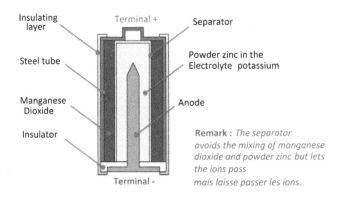

Figure 3.12. *Structure of an alkaline cell*

The voltage produced by this type of cell is the same as for saline cells. On the contrary, they display higher performances, as far as the lifetime and the current values provided are concerned.

3.3.3.3. *Other technologies*

Alkaline cells are those that are the most adapted to general use. We will note however that the so-called rechargeable cells are very quickly exploited as the surcharge of buying a charger is amortized at the end of around five recharge cycles when a rechargeable cell can handle around a hundred cycles. The cells only justify themselves for applications that consume very little energy and can function without being replaced for several months or years. Certain technologies are used for specific applications:

– lithium-based, non-rechargeable cells;

– silver oxide cells;

– zinc-air cells.

In the past, mercury cells have largely been used in gear that required a longer life span (a 10-year "shelf-life") but they have been banned because of mercury's toxicity. For the same reasons, the Nickel-Cadmium types of cells have eventually disappeared because of how dangerous cadmium is.

Generally speaking, certain technologies are interesting from the point of view of their calendar aging: we are speaking of "shelf life" to indicated that the cells end up degrading themselves without them being used in an electric circuit. The lithium based cells (for instance those in the CR2032 box) are particularly interesting since they can only provide rather low currents and are rather costly: we find this technology especially in power input circuits in the BIOS memory of computers in order to preserve information on their configuration.

The silver oxide cells are those that have the best energy density after the lithium technology. They are however costly and their main interest in relation to lithium cells is the absence of a risk of thermal runaway. There is, therefore no risk of combustion. For all of these reasons, we will find them in the fields of space and aeronautics (Saturn rocket, Apollo space module, etc.).

The zinc-air cells, although pretty rare, have the particularity of using oxygen from the environment for the oxidation reaction of the zinc contained in the cell. These cells only use a waterproof box and must be let in open air in order to function. We can see in Figure 3.13 an example of zinc-air commercialized by *Duracell* as the brand *Easy-Tab*: we can see here a new cell is equipped with a flap to use the cell. This type of cell is especially used for auditory prostheses because of their low cost coupled with a strong energy density.

Figure 3.13. *Example of zinc-air cell (Duracell Easy-Tab)*

3.3.4. *Cells from the viewpoint of the user*

From the point of view of the user, we can distinguish the cells according to the technology used, bearing on the characteristics and performances listed below:

– nominal voltage (in V);

– maximum available current (in A);

– capacity in A.h and consequently the energetic capacity;

– volume and mass;

– range of operating temperatures;

– OCV – *Open Circuit Voltage* and impedance depending on the state of the charge;

– shelf-life and usage constraints.

REMARK 3.2.– In numerous reference works, the capacity of a cell or a battery (expressed in A.h) is noted Q to mark the distinction from the capacitors and supercapacitors whose capacity (in Farads – F) is noted C. We are not going to make this distinction here because, even if this capacity does not have the same size (and approaches the notion of Coulombs charge – $C \equiv A.s$), we traditionally define the charge and discharge conditions by specifying the current depending on C and not Q (for instance $5C$, $2C$, C, $C/10$, etc.)[21].

---

21 More specifically, when a 60 A.h battery is charged at a C/5 current, we get a 12 A current and we will need 5 h to charge this battery if it was initially completely discharged.

Evidently, these characteristics are not completely independent of technological considerations and in particular of the use of standard boxes such as:

– the AAA format (cylinder of 10.5 mm in diameter and 44.5 mm in height, voltage: 1.5 V) known as IEC LR03 for alkaline cells;

– the AA format (cylinder of 14.5 mm in diameter and 50.5 mm in height, voltage: 1.5 V) known as IEC LR06 for alkaline cells;

– the C format (cylinder of 26.2 mm in diameter and 50 mm in height, voltage: 1.5 V) known as IEC LR14 for alkaline cells;

– the D format (cylinder of 34.2 mm in diameter and 61.5 mm in height, voltage: 1.5 V) known as IEC LR20 for alkaline cells;

– the 4.5 V format (a prism of 67 mm in height, 62 mm in size and 22 mm in thickness, voltage: 4.5 V) known as IEC 3LR12 for alkaline cells;

– the 9 V format (a prism of 48.5 mm in height, 26.5 mm in size and 17.5 mm in thickness, voltage: 9 V) known as IEC 6LR61 for alkaline cells;

– the CRxxyy formats that are lithium-based button cells with a nominal voltage of 3 V (open circuit voltage 3.6 V) in a flat cylinder format of xx mm in diameter and a thickness of yy/10 mm;

– the SRxx(x) formats that have silver oxide based button cells (nominal voltage of 1.55 V);

– the zinc-air cells whose references are 5, 10, 13, 312, 675 (but references of the type "ZAxx(x)" are also widely used). Their nominal voltage is 1.2 V.

There are several other cell formats but these ones are the most current ones and cover all of the most widespread technologies. The formats AAA, AA, C, D, 4.5 V and 9 V are also available in alkaline technology as well as saline technology. Let us note that rechargeable cells are also available for these formats, particularly with the NiMH technology which is introduced in section 3.3.8. The energetic capacity is another option in the choice of a cell for a given application in addition to the nominal voltage.

Evidently, the drop in voltage while in use (OCV variation + potentially variable impedance depending on temperature and cell aging) plays an important part in the good functioning of the system charged but the choice of a suitable capacity (in mA.h) is crucial. In order to be able to select the cells that work for an autonomous system, we can see in the table the values of typical capacities of different common cells. From an energy perspective, it suffices to say that the electrical power of "voltage × current" and, thus, that energy E_{pile} is the integral of this electrical power throughout a time period Δt. If we admit that the voltage V_{cell} and power I remain stable, it is then enough to describe:

$$E_{\text{cell}}[\text{J}] = V_{\text{cell}}.I.\Delta t = 3\,600 \times V_{\text{cell}}[\text{V}] \times \text{C}[\text{A.h}] \qquad [3.28]$$

where C is the capacity of the cell in A.h. In practice, this calculation is not as simple as the graph in Figure 3.14. Indeed, throughout the discharge of a cell, the voltage released varies: it drops strongly at the beginning and end of the discharge. The discharge is slower during the intermediary phase between 15 and 85 % of the discharge and the slope is highly dependent on technology. The tracks presented int his graph relies exclusively on a certain cell technology: alkaline technology which is, by far, the most widespread. It is compared to the technology of (rechargeable) batteries NiMH for which we can see that the voltage is lower (1.2 V instead of 1.5 V) but the characteristic of these batteries is a lot more constant depending on discharge (i.e. the charge used or released upon application) up to 2 A.h. We will therefore note that the energy stored is more important but also more easily exploitable because we dispose of source of voltage that is semi-constant on a very widespread functioning range. In the case of an alkaline cell, we will have to use high variation of voltage (-33% between the new cell and the cell that had released 2 Ah). We will note here the difference between this capacity and the one (typically) indicated in Table 3.6 for an AA alkaline cell that is supposed to have a capacity of 2 700 mA.h. However, we must mention that the values indicated in this table are heavily reliant on the conditions of the experiment such as *discharge current* and *cutoff voltage*. The latter is connected to the battery technology considered as in fact we are dealing with the practical value below which it is advisable not to go, particularly for batteries so as not to definitively alter their functioning for the future cycles (deep discharge). In the case of NiMH batteries and alkaline cells, these cut-off voltages remain quite close to each other since we consider that for the NiMH technology it is 1 V (as indicated in Figure 3.14) and it is of 0.8 or 0.9 V for alkaline cells (even if we have arbitrarily chosen the same threshold as for NiMH in this figure).

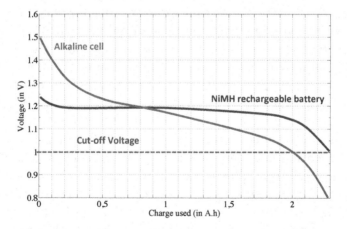

Figure 3.14. *Comparison of an AA alkaline cell discharge with an NiMH battery of the same size*

Format	Technology	Nominal voltage (in V)	Typical capacity (in mA.h)
AAA	Saline	1.5	540
AAA	Alkaline	1.5	1 200
AA	Saline	1.5	1 100
AA	Alkaline	1.5	2 700
C	Saline	1.5	3 800
C	Alkaline	1.5	8 000
D	Saline	1.5	8 000
D	Alkaline	1.5	12 000
4.5\,V	Saline	4.5	1 200
4.5\,V	Alkaline	4.5	6 100
9\,V	Saline	9	400
9\,V	Alkaline	9	565
CR1216	Lithium ($Li-MnO_2$)	3	25 (@ 0.1 mA)
CR2032	Lithium ($Li-MnO_2$)	3	225 (@ 0.2 mA)
SR41	Silver oxide	1.55	38 Å 45
ZA13 (or simply 13)	Zinc-Air	1.2	280

Table 3.6. *Typical capacities for different formats and different cell technologies*

The cell manufacturers propose results in a slightly different way in their datasheets, as we will see in Figure 3.15. This is an extract from the documentation of an AA Duracell cell from the "industrial" range (reference: ID1500). The latter is characterized for different discharge currents in terms of the voltage provided depending on time. The tracks provided all stop at $0.8\,V$ (which we will consider the factual cut-off voltage defined by the manufacturer) and we can thus see that the characteristic is not really proportional to the current released as the "simple" notion of A.h. capacity might make us think. When we look at the time needed for achieving "complete" discharge with a current of $50\,mA$, we find a duration between 55 and $60\,h$ whereas for a current of $1\,A$, the discharge time is no more than $1\,h$. We divide the discharge by at least 55 whereas the current is only multiplied by 20. This result, which is only apparently surprising, fully justifies that the cells are not characterized by an A.h. capacity but that this is the "norm" for batteries. This is obviously due to the fact that the energy provision is done with a certain electrochemical efficiency. We can note that at the energy level, the (purely) electric efficiency is in itself affected by the fact that the cell presents an internal resistance that the manufacturer indicates as having the value (120 mΩ @ 1kHz).

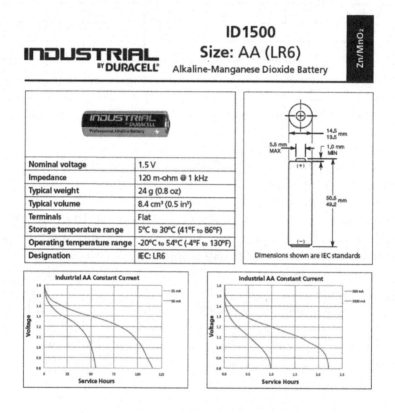

Figure 3.15. *Excerpt from an alkaline cell datasheet AA (Duracell Industrial ID1500)*

To make modeling this component even more complicated, the temperature also has an imact (as fir any other chemical reaction) on its behaviour. However, we can say that the behavior of a cell is better at a normal temperature (20 or 21°C) than at lower or very low temperature (0 or −10°C). This kind of characteristic must evidently be considered in the design of the electronic equipment meant to be used in extreme conditions.

Format	Nominal voltage	Nominal capacity C	Cut-off voltage	Nominal charge	Discharge rate
AAA	1.5 V	1150 mA.h	0.8 V	75Ω	$0.017C$
AA	1.5 V	2870 mA.h	0.8 V	75Ω	$0.007C$
C	1.5 V	7800 mA.h	0.8 V	39Ω	$0.005C$
D	1.5 V	17000 mA.h	0.8 V	39Ω	$0.0022C$
9 V	9 V	570 mA.h	4.8 V	620Ω	$0.025C$

Table 3.7. *Specifications of several alkaline cells (source: Panasonic)*

In fact, alkaline cells have a high energetic capacity compared to batteries. However, their characteristics link them to low power applications. In fact, it is not possible to provide them with high power cells because of their internal resistance. We can see in Table 3.7 the key (standard) characteristics in certain usual formats. These results show that the nominal charge resistance for a 1.5 V large-sized cell (format D) is 39Ω. Consequently, the current provided will be 38.5 mA whereas the capacity announced is 17 000 mA.h. In this nominal case we will have a discharge of $0.00226C \simeq C/442$! These results are consistent with the applications where the cells are used and their limits. A representative example of these limits is that of flashes in cameras, particularly the flashes used for reflex cameras. The current needed by the application (and also the voltage required for supplying the lamp) requires an electronic power converter associated to storage capacitors which will be able to provide the peak of the current required. In the case of full-power usage, the interval required between the two flashes will be quite long (more than 5 s measured for Nikon SB900 presented in Figure 3.16 to the left) because the recharging of capacitors from flash cells can only be done at limited current (four AA series cells, releasing a current of 20 mA below 6 V for a nominal load). In the case of intensive flash usage, this recycling time can be considered too long and is fully justified in the framework of a professional use of supplying blocks of flash equipped with batteries (Broncolor Move 1200 in Figure 3.16 on the right).

Figure 3.16. *Nikon SB900 flash (Left) and Boncolor Move 1200 flash and its battery-based power supply (Right)*

3.3.5. *History and principles of accumulators*

Accumulators, commonly known as batteries, are electrochemical devices capable of converting an energy contained in chemical reactives in electrical energy such as we have previously seen with cells. They have the ability to invert the process by storying electric energy by modifying the active chemical species at the interface between the electrodes and electrolyte. The first accumulator was the acid-lead battery, invented by *Gaston Planté* in 1859. Its internal structure is based on a series association of several elementary cells such as the one described in Figure 3.17.

The latter comprises lead electrodes submerged in sulfuric acid used as electrolyte. A separator (electronic insulator that lets ions circulate) has just come between the two electrodes and a vat (traditionally made of glass and nowadays plastic for obvious reasons of shock resistance) holds it together. Finally, the positive and negative electrodes can be differentiated as long as:

– the positive electrode has veins in between which tubes or lead oxide little razors or tubes are placed;

– the negative electrode has air cells filled with sponge lead[22].

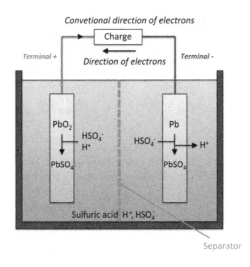

Figure 3.17. *Description of a lead-acid battery cell*

We can otherwise see the reaction at the level of the anode (oxidation) where the standard potential is −0.356 V:

$$Pb_{(s)} + HSO_{4(aq)}^{-} \rightarrow PbSO_{4(s)} + 2e^{-} + H^{+} \qquad [3.29]$$

whereas at the level of the cathode, we have the following reduction reaction (with a potential of +1.685 V):

$$PbO_{2(s)} + HSO_{4(aq)}^{-} + 3H^{+} + 2e^{-} \rightarrow PbSO_{4(s)} + 2H_2O \qquad [3.30]$$

---

22 Lead that has a large exchange surface with the electrolyte.

We will note that these reactions allow us to produce a theoretical voltage of $2.041\,V$ (in practice $2.1\,V$ for a fully charged battery). Besides, these reactions presented in the discharge mode of the battery, are perfectly reversible, as we can connect it to an accumulator (we only need to invert the direction of the arrows in the equations [3.29] and [3.30]).

A standard car battery ($12\,V$) has six series cells and is characterized by an extremely low internal resistance (of the order of $10\,m\Omega$). In these conditions, the short-circuit current of such a battery coudl reach $1.200\,A$. The fact is that the current required by the starter of a vehicle with a thermal engine is generally of several hundreds amperes. The lead battery proves to be well-adapted to this application, thus partially explaining the longevity of this technology.

3.3.6. *Comparing accumulator technologies*

We should now draw a comparison between battery technologies used nowadays in terms of energy density or specific energy (see Table 3.8). In fact, it is often the key parameter in choosing a battery (and even a battery technology) for an electric/electronic equipment with contained energy because the weight of the battery often represents a significant part of the total weight of the equipment (electric vehicle). It is however evident that the batteries are not comparable in terms of energy density with fossil energy: we need just remember that gas has a heating energy of $47.3\,MJ/kg$ (or $13.1\,kWh/kg$).

This comparative is not limited to storable energy density with a given technology. Indeed, the densities of materials being significantly different (11.35 for lead, 0.534 for lithium), a comparison of volume energies is also suggested. The existing link between the capacity in A.h of a battery (usual magnitude given by the manufacturer) and the energy stored going through knowing the voltage provided, the voltages released by an element are also given (when known) for each technology. We will note, in fact, that certain parameters are not indicated for certain technologies: one point of inquiry indicates that the magnitude is not known (either because of a lack of research – for instance for the Na-S technology – or because of a lack of hindsight on the technologies considered – for cycling and the self-discharge rates of certain batteries).

As this has been exposed with the Ragone chart at the beginning of this chapter, the energy stored is not always the predominant factor for the sizing of certain applications. We may indeed need to geta storing device for relatively low energies but before being provided in very short time: certain technologies are capable of significant energy peaks (Li-TiO) whereas others also have very limited capacities (Li-Air and Li-Po). Once the energy is stored, following the applications, we can keep a charging battery for a significant period of time without using it: we must then ask ourselves about the phenomenon of self-discharge that can be more or less acute depending on technologies: it can be a critical flaw for certain applications (such as the car[23]) whereas it is not as critical for others (camera batteries) that allow us to use NiMH batteries as these are mediocre in this regard.

Finally, the lifespan, throughout a number of cycles (charge/discharge) impacts on the usage cost of a given technology as well as the approval of the user:

– the unpleasantness of regularly changing the batteries;

– extra cost of buying replacement batteries.

Indeed, we consider in numerous applications that a satisfying lifespan for a battery must correspond to the

---

23 We will note that lead batteries, even if they are not the most highly performing at the moment, have a low self-discharge rate which is another factor in their favor and explains the fact that they have never been replaced in this field of application.

lifespan of the supplied product. This calculation can lead to a realistic solution for certain applications (phones and laptops[24], hybrid vehicles[25]) but for others (for instance electric vehicles) imposing criteria of adapted economical models (the location of batteries whose cost represents a considerable part of the global cost).

Finally, the last column of the table indicates for each technology that we are dealing with commercial batteries (production) or uniquely of prototypes developed in the lab (R&D). Let us note the presence of NiCd batteries that are no longer used because of the toxicity of cadmium but also a line reserved to alkaline cells (non-rechargeables[26]) As a comparison with (rechargeable) batteries.

There are no other aspects of the choice of technology of a battery featured in the Table 3.8. Certain technologies are more or less dangerous with risks of explosions or fires: this will be elaborated on a case by case basis in the section 3.3.8. Another point on the recyclability of batteries and the stocking up of main materials for their fabrication. This comes from the ecology and the eco-design but also from an economic viewpoint, this is a strategic aspect for the enterprises whose activity depends on the continuity of the production flow of batteries. But certain resources such as lithium could, in the future, become rare and lead to, if not a halt in the production, then at least a high increase in prices, making certain industrial activities non economically viable.

---

24 Considering the lifespans of one, two or three years for this type devices is debatable but clearly shown by certain manufacturers that put non-dismountable batteries in their products.

25 In the case of hybrid vehicles, the batteries are not used in the same demanding manner as in purely electric vehicles: their lifespan can then be quite long.

26 However, we will notice in the table the possibility of doing 25 to 500 cycles of charges/discharges with alkaline cells. In reality, there are dedicated chargers but it is important to note that the manufacturers of cells do not recommend this type of usage.

From this viewpoint, this difficulty joins that in the electronic or magnets industries at high performances with the stocking up of rare materials (samarium, dysprosium, neodymium, etc.).

Type	Energy density (in Wh/kg)	Volume energy (in Wh/l)	Voltage (in V)	Maximum Power (in W/kg)	Number of cycles	Self-discharge (in %/mois)	Statut
Acid lead	30 – 50	75 – 120	2.25	700	400 – 800	5	Production
Li-TiO	50 – 67	75 – 131	2.4	3 000	6 000	?	Production
Ni-MH	60 – 110	220 – 330	1.2	900	800 – 1 000	30	Production
Ni-Zn	90	280	1.6	1 000	200	≥ 20	Production
Na-S	100 – 110	?	?	?	?	?	Production
Li-Po	100 – 130	220 – 330	3.7	250	200 – 300	2	Production
Li-Fe-P	120 – 140	190 – 250	3.2	> 2 000	2 000	5	Production
Li-Ion	90 – 180	220 – 400	3.6	1 500	500 – 1 000	2	Production
Li-S	250	?	2.8	400	?	?	Production
Li-Air	1 500– 2 500	?	3.4	200	?	?	R&D
Ni-Cd	45 – 80	80 – 150	1.2	?	1 500– 2 000	≥ 20	Interdit (tox.)
Alkaline cell	80 – 160	?	1.5 – 1.65	?	25 – 500 ?	≤ 0.3	Production

Table 3.8. *Comparative of different battery technologies (one element)*

3.3.7. *Key quantities, definitions and vocabulary*

The field of batteries, at the interface between electric engineering and electro-chemistry, uses a specific vocabulary. We will limit ourselves here to recalling certain parameters that have already been introduced and introducing new ones but the latter are always oriented towards the use of batteries in the "world" of electric engineering.

From the viewpoint of electric modeling, the most rudimentary form of the electric scheme that is equivalent to a battery is a source of voltage in series with a certain resistance. Overall, it is a simple model equivalent to that of Thevenin:

– the voltage source is generally called OCV (*Open Circuit Voltage*);

– the series resistance is the internal resistance of the battery.

The difficulty of modeling these components lies (as for capacitors and supercapacitors) in the fact that the underlying (chemical) phenomena are strongly dependent on temperature. We will see in what follows that it has significant consequences at the monitoring level within a system. This monitoring aspect firstly has a bearing on the charge of the battery, which we note with C, and which is assessed in amperes.hours (A.h) and not in Coulomb (the equal unit that corresponds in fact to the Amperes/seconds) and turns out to be less practical. In fact the surveillance circuit of a battery (called BMS for *Battery Management System*) generally controls two large, different points:

– the charging state of the battery (that is, the reaction of the charging quantity remaining in the battery over the total capacity of the battery) which is traditionally called State-Of-Charge (SOC);

– the state of the health of the battery, noted State-Of-Health (SOH).

This parameter does not have, *a priori*, any precise definition. This is a positive factor on the part of the battery, which is representative of its conformity to the specs. We consider that a battery with an SOH of 100 % is fully functional (*a priori*, it is a new battery, whose behavior is in accordance with what is expected of it). When the performances start decaying, the SOH reduces and we can expect that a battery whose SOH drops at 0 % is no longer functional. In practice, we will set a threshold (which is usually not 0 %) below which the battery must be changed. Even if we can not give a unique definition to this magnitude, we can, however, draw up a list (rather complete but probably not exhaustive) of the measurable parameters of the battery that will be aggregated in the SOH, each with a particular weighting:

– internal resistance;

– effective capacity;

– OCV;

– self-discharge;

– number of charge/discharge cycles;

– temperature.

In certain cases, the SOH will have a far more restricted perimeter and will be limited, for example, to the assessment of the reducing capacity. We will complete the analysis of the state of the battery with an evaluation criterion called SOF (*State-Of-Function*) which will allow us to evaluate the capacity of the battery to ensure a given mission.

REMARK 3.3.– Although the SOC is the most common state of charge criterion, we also speak of discharge depth (noted Depth-Of-Discharge (DOD)) to represent the same notion. When the two quantities are expressed in %, we simply have $DOD[\%] = 100 - SOC[\%]$.

3.3.8. *Accumulator technologies*

3.3.8.1. *Lead batteries*

As it is indicated in the section 3.3.5, lead batteries are the most old ones and they are also the ones with the lowest energy density. They are however always used in the auto sector because of their perfect needs adjustment:

– they display a low self discharge (5 % per month);

– they are capable of releasing significant powers (700 W/kg);

– they are not very costly (compared to Li-Ion batteries for example).

They are not very dangerous to use. Nonetheless, there is a risk of explosion during the recharge phases (see [INR 12]), because during these phases, dihydrogenate is produced. When the gas mixes with the air around it, an ATEX (see Chapter 2) atmosphere is formed and the smallest spark can launch an explosion. We will note that in this situation, the destruction of the battery releases sulfuric acid equally dangerous for the persons and the materials (reaction with different materials, including metals such as bronze).

In terms of aging, in sealed batteries (so called "maintenance free" batteries) used in cars[27], we can expect a normal lifetime of five years but when the batteries are correctly kept with regular maintenance (control of the electrolyte density and adding distilled water), the lead-acid batteries (for instance used for energy storage associated to photovoltaic panels) can have lifespans of twenty years.

The main aging phenomena found in these batteries are:

– electrode oxidation (transformation of lead into lead oxide);

– geometric deforming of active matter (following numerous charge/discharge cycles during which the lead oxide PbO_2 is transformed into lead sulfate $PbSO_4$ and vice versa);

– the corrosion of electrodes by sulfuric acid whose concentration can finish by presenting an ingredient[28];

---

27 AGM (*Absorbed Glass Mat* batteries), that is with an electrolyte stored in "blotting paper".

28 This problem, known as *acid stratification*, is essentially present when the electrolyte is free in liquid state, and not (or little) in the case of AGM batteries.

– sulphating (i.e. crystallization of lead sulphate in the electrode) when the battery is maintained in discharge state (or poorly charged) during a long time.

Generally speaking, the aging of batteries translates as:

– low capacity;

– increase in internal resistance;

– reducing open circuit voltage (OCV).

The latter can be significant and representative for a short circuit of one or several cells. This problem is quite rare on modern batteries but can finally happen if conducive elements settle at the bottom of the vat and come into contact despite the separator.

Regarding sulfating, which is a common problem in lead batteries, there are regeneration methods in place for increasing their lifespan and countering this phenomenon. These methods are of two kinds:

– electrical: with an electronic re-generator sending current pulses at specific frequencies in the battery for eliminating sulfate crystals;

– chemical: with cleaning agents.

From the point of view of the user, we essentially qualify in terms of nominal voltage and A.h capacity. In the case of automobile batteries, we mainly use 12 V batteries and the standard capacity for a large Sedan can reach 100 A.h. We can see in Table 3.9 the characteristics of a Varta battery (range: Silver Dynamic AGM) whose photo is shown in Figure 3.18. In literature we find usage recommendations such as reasonable charges and discharges not going over 4 C (C being the capacity of the battery in A.h)[29].

Reference	595901085D852
Gamme	Silver dynamic AGM
Battery capacity [Ah]	95
NE cold testing current [A]	850
Height [mm]	190
Size [mm]	175
Length [mm]	353
Voltage [V]	12

Table 3.9. *Technical specs of a lead-acid battery for an automobile (Varta)*

Figure 3.18. *Image of a lead-acid battery of 12 V (reference: Varta 595901085D852)*

Finally, we note that hybrid lead batteries associating classical batteries to supercapacitors have been developed by an Australian company called Ultrabattery® (www.ultrabattery.com). These batteries allow us to offer the energetic capacity of a battery while allowing us to provide power peaks due to the supercapacitors without adding

---

29 We will note that this recommendation is not followed in the case of car batteries (less than 400 A for the battery in Table 3.9 whereas the ignition of a car often consumes more than 600 A and can be used very often in vehicles equipped with a Start-Stop function).

electronic power converters between these elements that are combined in a single component, as shown in Figure 3.19.

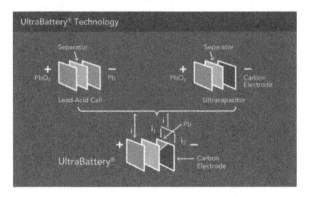

Figure 3.19. *Diagram of the Ultrabattery principle*®
(source: www.ultrabattery.com)

From the viewpoint of the user, bringing this technology has a significant imapct on increasing the lifespan of the battery (see Figure 3.20) (undoubtedly by the fact that the power peaks delivered no longer put a strain on the battery but on the supercapacitor) and higher efficiency (ratio between energy recovered in discharge) and energy injected during charge.

3.3.8.2. *Ni-MH Batteries*

Ni-MH (or nickel-metallic hydride) batteries are made of:

– an electrode positive in nickel oxy-hydroxide;

– from a negative electrode with an alloy of lanthanum[30] and nickel ($LaNi_5$) is a metallic hydrate (a metal capable of storing hydrogen);

– potassium (KOH) that works as an electrolyte.

---

30 Lanthanum is rare earth mineral. It is a soft, gray, ductile metal that can be cut with the knife at room temperature.

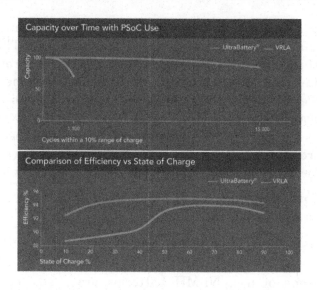

Figure 3.20. *Influence of the "Ultrabattery® technology" (source: www.ultrabattery.com). For a color version of the figure, see www.iste.co.uk/patin/power5.zip*

The voltage provided by a Ni-MH cell is 1.2 V. This technology has replaced the Ni-Cud technology which was banned in Europe following a decision made in 2006. This ban is due to the presence of cadmium, which is toxic for humans (causes renal problems and increases blood pressure) but also because of the environment even in very low concentration.

The Ni-MH technology is more performing than the Ni-Cd technology on certain points (mass energies and volume energy) as it is shown in the Table 3.8 but it fares worse in regards to lifespan (number of charge/discharge cycles) but also and above all regarding self-discharge it is higher or equal to 30 % per month. In fact, it is the major problem of "standard" Ni-MH batteries because 10 at 15 % of the charge is lost in the first 24 h following the charge. The discharge can then be of 15% per month. The discharge indicated in the Table 3.8 can therefore impact the first month after the charge but it is worth noting that this value is significantly higher than for other battery technologies presented. This flaw has pushed industries to develop batteries that improve this parameter: the first manufacturer who has implemented a Ni-MH technology with a low discharge (Low Self-Discharge (LSD)) was Sanyo (powerly a part of Panasonic) in 2005. These batteries are sold pre-charged and ready to use and then when the improving of specs is due to the use of a more performing separator between the two electrodes. This translates as:

– a significant increase in the number of charge/discharge cycles (2, 100);

– a retention of 70 to 85 % of the initial charge after a year.

In terms of use, Ni-MH batteries are used in various electronic devices: this technology is used to manufacture rechargeable cells in the AA and AAA formats, for instance. We will however note that this technology is not only limited to low power applications because we can also find it in hybrid vehicles and in particular in the most well-known of them all: Toyota Prius. We can see in Figure 3.21 the photographs of the two extremes presented: an AA rechargeable cell and the battery of the Toyota Prius II (both with Ni-MH technology).

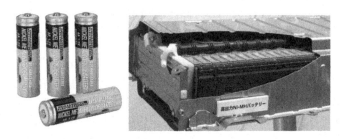

Figure 3.21. *Two types of Ni-MH battery (sources: Chicago Electric (Left), Toyota Prius II (Right))*

3.3.8.3. *Lithium-based batteries*

Lithium-based batteries are considered to be the most modern and the most effective amongst all available technologies. In fact, the first lithium battery was proposed in the 1970s[31]. Amongst these, we can consider the Li-Ion batteries as the most widespread, not only for supplying "wide audience" electric devices (mobile phones, laptops, etc.) but also for more and more powerful applications (electric vehicles such as Tesla). Its performance in terms of mass/volume energies, mass power, lifespan and self-discharge make this an attractive polyvalent battery. In fact, Table 3.8 does not show any fault with this technology[32]. Nevertheless, these batteries display a major risk: if they are misused, a thermal runaway could lead to inflaming the lithium contained in the battery and risks to induce a catastrophic fault in the device. We can mention here the example of the Boeing 787 of Japan Airlines whose Auxiliary Power Unit (APU) caught fire on the 7 of January 2013 (see Figure 3.22).

Lithium is therefore both an interesting product (at the level of the energy stored and its mass, given the density of lithium[33] is 0.534) and problematic because it reacts strongly[34] when in contact with the oxygen from air or water. In case of thermal runaway, the combustion of the lithium battery is thus incontrollable because even if we deprive it of oxygen, it will be capable of producing it to maintain its combustion [HEM 06] until the complete consumption of the combustible matter. Setting aside the obvious inconveniences of a fire[35], the results of the combustion of a lithium battery

---

31 The Ni-MH batteries date back from 1960 and their ancestors Ni-Cd as far as the end of the 19th Century.

32 Even if other technologies can be better in one specific point.

33 This is the metal with the smallest density.

34 This is a classic euphemism in chemistry!

35 Yet another euphemism.

turn out to be toxic: we can mention hydrogen fluoride (HF, because of the presence of fluorine in lithium salt).

Figure 3.22. *Li-Ion Battery destroyed by the fire of the APU of the Boeing 787 of Japan Airlines JA829J (source: National Transportation Safety Board)*

As the name indicates, in a Li-Ion battery, the lithium is present in an ionized state (Li^+) traveling between the two electrodes of the cell, one made of a metal oxide (cobalt or manganese dioxide) and the other of graphite. In this kind of battery, the electrolyte used between these two electrodes is considered aprotic: this means that it does not either accept or release protons, which invites interactions with the electrodes (that destroy them). Concretely speaking, this electrolyte is made of lithium hexafluorophosphate ($LiPF_6$) in a solution of an organic solvent (for instance diethyl carbonate – $OC(OCH_2CH_3)_2$). On the basis of this physical description (completed by the diagram in Figure 3.23), we can draw a list of faults associated to their causes which can damage a Li-Ion battery because of poor use or normal aging:

– the formation of metallic lithium dendrites at the surface of the negative electrode in case of surrcharge (especially at low temperature since high currents capability is reduced);

– thermal runaway at high temperatures or in case of too high heating of the battery (the maximum acceptable current is also reduced at high temperature).

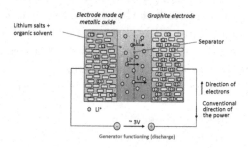

Figure 3.23. *Structure of a Li-Ion battery*

Temperature is thus a significant factor that needs to be considered in order to properly manage the use of a Li-Ion battery. The electric short-circuit is obviously another phenomenon that triggers the catastrophic failure of the battery and all of the protections required must be implemented to prevent this situation. The dendrites that we mentioned above can lead to this type of failure even inside of the battery: in fact, these are small points (metallic crystalline formations) that pierce the separator and lead to a localized short-circuit of the battery[36]. As a general rule, this (localized) short-circuit is too weak to bear the current that will circulate: it is instantly destroyed. However, when this phenomenon is not repeated, it leads to a progressive destruction of the separator and will lead to short-circuits that are significantly more critical for the cell as a whole. Even before reaching a major failure of the battery, the lithium brought to a metallic state becomes unusable for the charge (and discharge) of the battery. What is more, the

---

36 Let us remember that the stacking of electrodes and the separator (impregnated with electrolyte and lithium salt) can have a thickness as low as 10 μm (100 μm more traditionally).

surface of the metalized electrode becomes inoperable: we then see a loss in capacity.

The conclusion that can be drawn about Li-Ion batteries is that it presents use risks but this is equally the case of lead batteries that can emit di-hydrogen in case of overcharge (water electrolysis). It is therefore worth monitoring and controlling the batteries within systems that are meant to use them: this aspect will be discussed in section 3.4.

It is worth making these remarks more nuanced since there are different variants of Li-Ion batteries, each with its specificities. What has been previously mentioned essentially corresponds to batteries of $LiCoO_2$ type (Cobalt Lithium-dioxyde[37]). Other technologies such as:

– Lithium-manganese dioxide – $LiMnO_2$ (called LMO);

– Lithium-Iron Phosphate – $LiFePO_4$ (called LFP);

are well known for being less sensitive to thermal runaway. On the contrary, they do not perform as well from the viewpoint of energy density.

There are four other lithium battery technologies mentioned in the comparative Table 3.8:

– Li-Po batteries (lithium-polymer);

– Li-S batteries (lithium-sulfur);

– Li-TiO batteries (lithium-titanate);

– Li-air batteries.

First of all we will discuss the most classic technology (Li-Po) which is commonly used in widely available applications

_____

37 To finish the portrait of battery toxicity, cobalt is suspected of causing cancer.

(for instance common batteries in radio operated vehicles but also on bikes with electric assistance or even in certain vehicles). They have the advantage of being:

– slim and light;

– potentially flexible;

– safer than classic Li-Ion batteries.

On the other hand they are more costly and do not perform as well as the latter. They are always subjected to strict charging rules but they react better to overcharging and since the polymer is a solid electrolyte they are not as sensitive to faults as the classic Li-Ion batteries[38], using a liquid electrolyte.

Lithium sulphur batteries are modern batteries whose performance is promising but, as is shown by Table 3.8, a certain number of key characteristics are missing (in particular the number of charge/discharge cycles). As the website of Sion Power company shows (www.sionpower.com), there seems to be a beginning in the industrialization of this technology. One of the strong points put forth by this technology is a reduction in costs compared to the classic Li-Ion technology.

The Li-Tio technology was implemented by Toshiba and is used for mobile IT hardware and brings important advantages to this type of applications, such as a longer lifespan and a quick recharge. On the other hand, it offers an energy density three times lower than the Li-Ion technology.

Finally, the Li-Air technology is, as mentioned in Table 3.8, still in a developing stage but seems to offer considerable energy storing potential (around 10 times the energy density of other technologies) but there are still a lot of unknown

---

38 Let us note that the Li-Po batteies are in fact a variant of classic Li-Ion batteries and not a separate family.

factors in its global performance. The gain in energy density comes, partly, from the fact that one of the active materials (the air, or, more precisely, the di-oxygen contained in the air) is not included in the mass of the battery: we find in this a property of zinc-air cells that we have presented previously. On the other hand, the technical difficulties such as the corrosion of electrodes that is needed to filter the air brought to the battery are yet to be solved to turn this lab prototype into a viable industrial solution. At present, BMW and Toyota have joined forces to develop (amongst other things) Li-Air batteries for electric and hybrid vehicles.

3.3.8.4. *Another Lithium technology: LMP*

Another technology that is not mentioned in Table 3.8 is also used industrially: the "Lithium-Metal-Polymer" (LMP) technology. It is characterized by a long lifespan and thus the best option for electric vehicles. It is the technology used by BlueCar and developed by the Bollore group (see Figure 3.24). In fact, this technology is also developed in this group via the BatsCap company created in 2001.

Figure 3.24. *BlueCar (used particularly through the Auto'Lib service in Parisian areas)*

In this technology, the anode is made of metallic lithium sheets whereas the cathode is obtained from vanadium oxide, carbon and polymers. The electrolyte is made with polyoxiethylene (a solvent) mixed with lithium salt. The structure of this battery is in fact, entirely solid: there is no liquid or pasty electrolyte and so, there is no leakage risk. This technology is thus robust, along with having a high durability (in the number of charge/discharge cycles): the batteries that equip the BlueCar are known for having a lifespan of 400,000 km. They present a major inconvenience as far as their use is concerned: to function, they must be brought to 60°C (for a use in charge or in discharge) and to a minimum of 40°C during the maintenance phases of the charging state (i.e. *floating*). It is therefore worth evaluating the energetic cost of the temperature maintenance of batteries in order to analyze the global expense for the predicted use of these batteries.

3.4. Monitoring the charge states and the health state of components

3.4.1. *Batteries*

3.4.1.1. *OCV characteristics*

In practice, the measure of the state of charge of the batteries largely rests on the OCV characteristic of the battery. We remember that the OCV of a battery is its voltage in vacuum: therefore, it can only really be measured precisely when the battery does not provide any current[39]. To make a correct measurement, we must wait long enough after the use of the battery so that the relaxation phenomenon inside the battery has stopped for the voltage to be actually stabilized. Based on these preliminary observations, we can now look

---

[39] If it does provide current, we will talk about estimation and this approach will only be discussed in the section 3.4.1.3.

into the voltage released by a battery depending on its state of charge (SOC). Generally speaking, we find characteristics analogous to those of a cell (as seen in Figure 3.14). In this diagram, we already have a characteristic of the battery (Ni-MH rechargeable cell) presented in comparison to that of an alkaline cell and we could also note that this characteristic was semi-horizontal on a wide range of battery functioning[40]. If from the viewpoint of the user, the battery makes an excellent source of voltage, this poses a very concrete problem for the estimation of the state of charge, which is illustrated by the graphs in Figure 3.25.

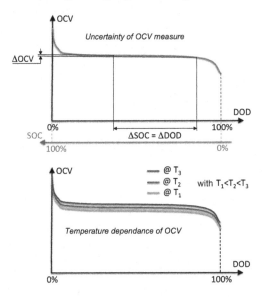

Figure 3.25. *The OCV characteristic depending on SOC and the measuring difficulty associated with it*

As we can see from the graph at the top, the smallest uncertainty in the measure of the voltage can cause a serious error in the estimation of the SOC (or here the depth of

---

40 This is actually the case of LiFePO$_4$ batteries.

discharge (DOD)). If this error can be reduced by taking care of the measuring circuit (we can hope to take the precise measurements down to the millivolt[41]), we can also note that the OCV curve depends on the temperature (see the graph on the bottom) and that for a given state of charge, the voltage can vary (more or less depending on the technology) and also depending on temperature, it also having an impact on the quality of the charge estimation if it is not correctly taken into account. Finally, the battery presents an OCV that varies in the time of:

– its aging;

– its previous mode of functioning (charge/discharge).

This latter point is particularly critical for certain technologies (Ni-MH but also Li-Ion for example) because the OCV curve can present an hysteresis that makes the estimation of the state of charge completely erroneous if we consider the in charge characteristic to follow discharge or the other way around. This is clearly illustrated in the curve in Figure 3.26 taken from [ROS 11].

As for aging, it must be integrated in the estimation of the state of charge by an identification of the parameters others than OCV (battery impedance) but also by estimating its real capacity. We will not speak about it in detail here as it falls outside of the scope of this book and we will limit ourselves to presenting an overall view of the "monitoring" function of a BMS in the section 3.4.1.3.

3.4.1.2. *Coulomb counting*

The OCV measurement allows us to estimate the state of charge starting from the moment where the OCV

---

41 On the condition of managing the rejection of the common mode well (see Chapter 1).

characteristic was identified. During its functioning, the OCV is no longer accessible given that the voltage V_{batt} at the terminals of the battery will be equal to:

$$V_{batt} = OCV - Z_{batt} \cdot I_{batt} \qquad [3.31]$$

where Z_{batt} is the impedance of the battery and I_{batt} is the current released by the battery[42]. On the basis of a precise measuring of the voltage and the current as well as an identification of the impedance of the battery, we are, theoretically, capable of estimating in each instant of the functioning of the system – the OCV of the battery. In practice, this task is complex because:

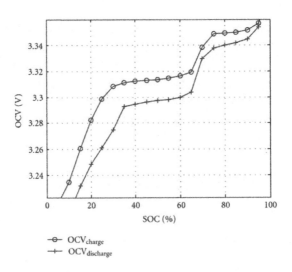

Figure 3.26. *OCV characteristic depending on the SOC following the charges and discharges at* 0.5 *C (measures after 3h in idle mode) – source: [ROS 11]*

---

[42] We place ourselves wihin the framework of the functioning of the generator (so in discharge).

– the measures required must be precise;

– a system that includes electronic power converters is noisy;

– the temperature of the battery impacts on its OCV and its Z_{batt} impedance.

For all of these reasons, a simpler method is often preferred. Estimating the SOC based on an initial value SOC_0 by integrating the current released (or absorbed) by using the following relation:

$$\text{SOC}(t) = \text{SOC}_0 - \frac{1}{C} \int_{t_0}^{t} I_{\text{batt}}(\tau).d\tau \qquad [3.32]$$

where t is noted in hours if the capacity C is expressed in A.h. This calculation is known as *coulomb counting* since it literally boils down to calculating the charge released by the battery. Theoretically, this functions perfectly but in practice it means a perfect measuring of the current. It is obviously not possible in reality and it leads to obtaining a result that strays away, more or less quickly, from the real SOC of the battery with all the risks that it entails:

– the possibility of a deep battery discharge when it releases a current ($I_{\text{batt}} > 0$);

– the possibility of an overcharge of the battery when it is in recharge ($I_{\text{batt}} < 0$).

Finally, in the more "rudimentary" equipments, we use this technique in association with a rescaling of the SOC at each stop. In fact, we remember that when $I_{\text{batt}} = 0$ for long enough a time, the voltage becomes again equal to the OCV and we can, again, estimate the SOC thanks to the OCV (SOC) characteristic of the battery.

3.4.1.3. *Overview*

The management of batteries is ensured in a device called Battery Management System (BMS). This can be more or less sophisticated according to the need for precision upon knowing the state of charge and the health state of a battery pack. A general description proposition of such a function is presented in Figure 3.27. We can see that the BMS uses, as we have said previously, an estimation of the OCV to initialize the initial state of charge of the battery. This can be done while functioning if we know (or identify) the impedance of the battery or, more easily, in vacuum (I_{batt}), once the voltage provided by the battery is stabilized.

3.4.2. *Dedicated integrated circuits*

The manufacturers of semi-conductors have developed a range of components dedicated to *power management*. Among these, we find integrated electronic power converters (drivers and voltage/current regulation circuits) but also circuits dedicated to the management of sources and in particular, *battery management*. The most current circuits are those allowing to charge the batteries (chargers that are initially integrated or at least circuits allowing the control of electronic power converters dedicated to charging a battery[43]): several manufacturers propose products dedicated to this function, as the one shown in section 3.5.3.6. In the case of BMS, the offer is more limited. The "BMS" function can be seen as a conglomerate of functions, which includes:

– detection of state of charge;

– protection;

– balancing.

---

43 The "all integrated" solutions being obviously limited to low power applications.

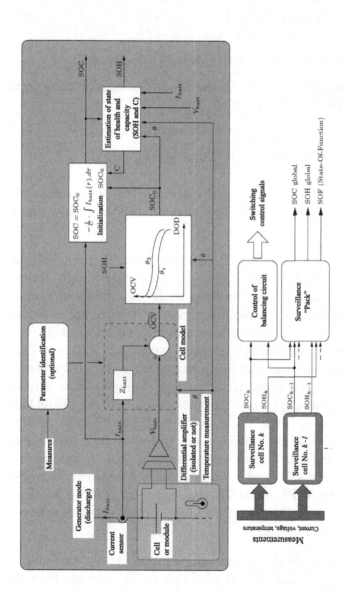

Figure 3.27. *Simplified structure of a BMS (battery management system)*

The detection of the state of charge (i.e. *fuel gauges*) is, without a doubt, the most important aspect (at least in comparison to the volume of components sold) and among the manufacturers that propose this type of function, we find:

– Texas Instruments (with the bq27xxx circuits);

– Maxim Integrated (with the ModelGauges, ModelGauges m3 and FuelPacks ranges);

– Linear Technology (LTC4150 and LTC294x(-1));

– ST Microelectronics (GG25L and STC31xx).

Circuits are dedicated to the monitoring of batteries (for protection purposes). To do this, they ensure the monitoring of temperature.

The balancing of a function is more rare. Perhaps because it is only implemented in medium and high power applications which are, in fact, more rare than the other applications such as mobile devices (telephones, tablets, laptops, etc.). One particular company stands out in this application field: Intersil, who proposes battery pack management circuits including balancing cells with a strong adaptability compared with voltage ranges that can be processed. Indeed, these circuits such as the ISL94212 (managing, on its own, six to twelve cells) can be cascaded to manage up to 168 cells. We can see in Figure 3.28 (to the left) the basic diagram for the connection of this circuit, with a battery on one side and a microprocessor or microcontroller on the other (SPI bus). To the right, a cascade of three evaluation cards (ISL94212EVZ) is implemented in order to manage a 36-cell pack.

We will note that the charging function is not included in high power applications, which is perfectly normal given the impossibility of integrating a power converter in a dedicated chip. Indeed, monitoring circuits only contain low power circuits (the balancing function set aside). In contrast, at low power, it is common to use general circuits that ensure, simultaneously:

– the charging of the battery;

– the monitoring of the battery;

– the energy supply of the application;

– the selection of the main source between the utility grid (or more traditionally a USB port[44]) and for the battery.

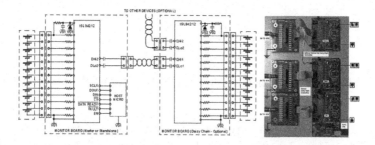

Figure 3.28. *Intersil ISL94212 (cabling diagram to the left and cascade of three evaluation cards to the right)*

In this respect, there are certain microcontroller manufacturers (such as Microchip or Texas Instruments) that propose such functions. As an example, we can mention the case of the Microchip MCP73871 circuit which is particularly simple to implement, as is shown by the application diagram in Figure 3.29.

3.4.3. *The case of capacitors and supercapacitors*

In the case of capacitors, a similar approach to the one used for batteries may be used. We will note however that the equivalent of the OCV is more simple to determine and that the characteristic of such a component is the simplest

---

44 Either way, for all "IT" applications.

possible version of a battery because it is linear. We will therefore not experience the same weather difficulties as for a LiFePO$_4$ battery for which the OCV is semi-constant throughout the majority of the SOC variation range.

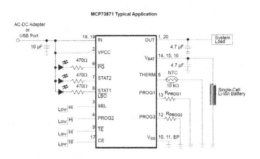

Figure 3.29. *Application diagram of the Microchip MCP73871 circuit*

In the case of capacitors the notions of SOC and especially OCV do not make sense and the monitoring of a capacitor will be limited whereas a real time identification of the parameters of one of the models in Figure 3.3. The major difficulty is to integrate in an extended Kalman filtering (for instance) the identification of the states of the model improved with resistances and (ideal) capacitors and then of finding the good solicitations to apply in the commands of a converter to allow an *in situ* analysis. Either way, the impact of the temperature cannot be neglected to hope to obtain correct information on the state of the components.

In relevant literature, we find "rudimentary" equipments such as those in [VOG 11] to analyze the state of a capacitor via a simple assessment of its ESR. Once the temperature has been identified, confronting the expected value with the measured one can easily help generate an alarm in a command body in case the difference is too significant. Generally speaking, all of the parameter monitoring approaches translate into a similar method. However, the aging of the capacitors can prove more complex and can make the use of a synthetic parameter (such as the ESR) too reducing to "see" the aging through life and risks only giving a sensible indication when the component is practically at the end of its life[45].

3.5. Circuits associated with storing systems

3.5.1. Decoupling capacitors

The use of decoupling capacitors has already been addressed in the case of a three phase inverter in Chapter 6 of Volume 2 [PAT 15b]. The electrolytic capacitors are generally sized on the basis of the RMS current that they will have to process and not on their capacitance. This was demonstrated in [PAT 10] at the most critical operating points for an entire range of electrolytic capacitors: the voltage ripple is not generally the sizing criterion (except for very low switching frequencies) and only the RMS value of the current can impose the use of supplementary capacitors so as not to overburden the components.

3.5.1.1. Efficient power

In practice, the sizing of decoupling capacitors for choppers is relatively easy[46]. This is not the case of three-phase inverters but also for single-phase ones as different Pulse Width Modulation (PWM) strategies are applicable to reach the same functional result. In fact, the more complex the converter, the higher the number of degrees of freedom in their command. Theoretically, there is an infinite number of PWM

---

45 It is obvious that this is not what we look for when we implement a monitoring device.

46 Unless optimizing the sizing in regards to a specific mission profile allowing us to play on their thermal inertia, the temperature being a key factor in their aging.

strategies that we can apply to a classic three-phase inverter and, in practice, several variants are indeed used:

- sinusoidal modulation (SPWM);

- Space Vector Modulation (SVPWM);

- discontinuous modulation (several variants xDPWM);

- double carrier modulation DCPWM;

- Δ/Σ PWM;

- stochastic PWM;

- pre-calculated PWM;

- full wave modulation;

- hysteresis (band) control;

- etc.

All of these strategies have a different bearing on the decoupling capacitor(s) for a functioning point given (i.e. "charge"). A rather complete study on this subject was carried in the electromechanics lab in Compiegne via the following theses: [HOB 05] in 2005 and [NGU 11] in 2011. The standard strategy used as reference was the SVPWM for which we know, thanks to [DAH 96], the analytical expression of RMS value of the AC component of the current absorbed by the inverter on the DC bus[47]:

$$RMS\left(I_c\right) = I_{\max}\sqrt{\frac{\sqrt{3}m}{4\pi} + \left(\frac{\sqrt{3}m}{\pi} - \frac{9m^2}{16}\right)\cos^2\varphi} \quad [3.33]$$

where $I_{\max}$ is the amplitude of the currents in the load, m is the modulation index (such as the amplitude of fundamental

47 Ideally, all of this AC power circulates in the decoupling capacitors and only the average value is picked up on the source.

phase voltages provided to the load which is equal to $V_1 = m.V_{dc}/2$[48]) and $\cos\varphi$ being the power factor of the load.

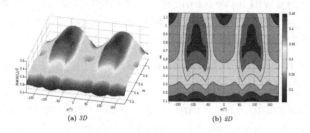

(a) 3D   (b) 2D

Figure 3.30. Cartography $\frac{RMS(I_c)}{I_{max}} = f(\varphi, m)$ obtained for the Uni-DCPWM strategy. For a color version of the figure, see www.iste.co.uk/patin/power5.zip

This result, already presented in Chapter 6 of Volume 2 [PAT 15b], allows us to evaluate other strategies at I_{max} given for all the values m and φ given (between 0 and 1.15 for one and $-180°$ and $+180°$ for the other). In [NGU 11], a strategy that has been called Uni-DCPWM (for *Unified-Double Carrier Pulse Width Modulation*) has allowed us to obtain the cartography $\frac{RMS(I_c)}{I_{max}} = f(\varphi, m)$ presented in Figure 3.30 and consequently a reduction coefficient in relation to the SVPWM whose cartography in function m and that of φ is presented in Figure 3.31.

With this figure, it can be seen that the currents in capacitors can be reduced with over 40 % to the high power factor ($\varphi = 0$) for a modulation index m close to 0.7. Generally speaking, we notice that this strategy is very favorable for capacitors for all the realistic power factors, from the engine ($\varphi = 0° \pm 50°$) to the generator ($\varphi = \pm180° \pm 50°$). This gain of 40 % at the level of the efficient power is translated into the given resistance (the ESR of capacitors), by a reduction of losses of 64 %. This result is thus remarkable but it must be nuanced by the fact that if the capacitors heat less, their

---

48 V_{dc} being the voltage of the continuous bus.

ESR will be lower. However, the main goal is to achieve the lowest temperature: capacitors that have a lower internal temperature will have a longer lifespan. The degree of freedom offered[49] to the developer of the electronic power converter (including the decoupling capacitors) will be:

– to reduce the volume of capacitors for a level of stress maintained;

– to maintain the same volume of capacitors and thus of decreasing their stress (and therefore lengthening their lifespan).

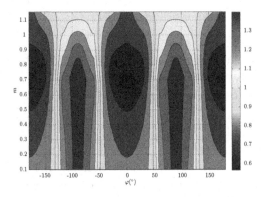

Figure 3.31. *Cartography comparison* $\frac{RMS(I_c)_{\text{Uni-DCPWM}}}{RMS(I_c)_{\text{SVPWM}}} = g\,(\varphi, m)$. *For a color version of the figure, see www.iste.co.uk/patin/power5.zip*

Even if the analysis of this PWM technique is outside of the scope of this chapter and work, the price to pay by applying this strategy resides in the increase of the distortion of voltages provided to the charge and therefore an increase in the distortion of the voltages provided to the charge and thus an increase in the distortion of currents. Indeed, despite the fact that the current waves remain visually moderate (in the oscilloscope) the spectral analysis shows a larger

---

49 By going from a classic SVPWM to a Uni-DCPWM strategy.

harmonic richness, that might easily excite the resonance modes of the machine thus supplied, significantly increasing the acoustic nuances. To prevent this type of problem it is worth increasing the frequency of the converter breakdown.

The spectral aspect of the decoupling problem must not be omitted in a sizing (of the decoupling of the DC bus). We can already mention the "panachage" of technologies and of the sizes of capacitors in order to offer a low impedance on a frequency band large enough so as to manage all of the frequency components induced by commutation:

– electrolytic capacitors for managing the first multiples of the decoupling frequency (and potentially the harmonics of the fundamental in case of overmodulation or of full wave modulation and other low-frequency components such as the multiples of the voltage of the (utility) grid in case of connection to a rectifier in an industrial variable speed drive);

– non-polarized capacitors for managing the highest frequencies up to the ones induced by the commutation fronts[50];

– the size of the capacitors can even play a role among non-polarized capacitors when the frequencies that need filtering are very high and the series inductance of the capacitor can ruin the result: we prefer more and more SMD capacitors in these cases. Very clear designing rules are otherwise presented regarding this Xilinx application note [XAP 05] for the decoupling of the alimentations of a FPGA.

It is therefore worth analyzing the quantity of current that will be process by this or that capacitor technology depending on its limited frequency (we can consider that it is its own frequency $f_0 = \frac{1}{2\pi\sqrt{LC}}$ where C is the value of the capacity

---

50 These aspects are mentioned in Volume 4 [PAT 15d] dedicated to electromagnetic compatibility (CEM).

in Farads of the capacitor whereas L is its equivalent series inductance, powerly noted with ESL – potentially increased compared to that of the cabling in relation to the power switch that we wish to decouple). An efficient way of dealing with this problem is to represent, for a strategy and functioning point given of the charge and for each frequency f, the *partial RMS value of the current* below this frequency f. In order to obtain these most general tracks possible, it is reasonable to normalize the axes:

– for the abscissa axis we will use a normalized frequency $\nu = f/F_d$ (where F_d is the decoupling frequency used by the PWM controller);

– for the ordinates axis we will use a normalized partial RMS current $i^{\text{part}}_{c-RMS} = RMS\left(I^{\text{part}}_c\right)/RMS\left(I_c\right)$: therefore, when $\nu \to \infty$, this current i^{part}_{c-RMS} tends towards 1 (or 100 %);

For illustrative purposes, we can see in Figure 3.32 the curve of this current for a SVPWM strategy with a modulation index m variating from 0.2 to 1.15 ($2/\sqrt{3}$ – this is the linear modulation limit[51]).

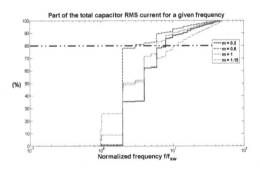

Figure 3.32. *Normalized partially RMS current i^{part}_{c-RMS} depending on the normalized frequency ν. For a color version of the figure, see www.iste.co.uk/patin/power5.zip*

---

51 See Chapter 2 of Volume 2 [PAT 15b].

In this figure, we can easily evaluate up to what frequency (knowing the frequency of the breakdown) 80% of the total RMS current is divided (from the AC component of the DC bus current – therefore, theoretically, from the current in capacitors). We realize that for the SVPWM strategy, the modulation index plays an important role in the spectral distribution of the current:

– for $m = 0.6$ for example, 80% of current is below three times the frequency of the breakdown;

– whereas for the inferior and superior modulations indexes, we must go significantly higher (more than ten times $m = 1.15$).

We will, however, note that this figure must be used in conjunction with that evaluating the global RMS current as we then realize that for $m = 0.6$ it is very high (at any rate, (for a given $I_{\max}$) as it has a tendency to decrease for the low and high modulation indexes (in particular for $m = 1.15$).

3.5.2. *Balancing circuits*

3.5.2.1. *Objectives*

Balancing circuits are used for supercapacitors as well as for batteries. In fact, in their passive form, they are used for a wide range of components that we do not expect, when we associate it in a series, to divide their total voltage equally. Similarly, the current reaching a node on supposedly identical components placed in parallel does not generally divide in equal currents in each line. These imbalances can generate catastrophic failures in a system: in fact, if a component is used beyond its capacities (in voltage as much as in current) a cascade of failures may well take place.

For instance, if a supercapacitor placed in the same line as other series supercapacitors witnesses too high a voltage at its terminals, an electrolysis of the solvent will take place and gas will be generated. The destruction of the component will soon follow. From then on, two situations can take place:

– either the supercapacitor in failure becomes an open circuit: the entire line is then unusable. In contrast, the other supercapacitors remain intact;

– or the failing supercapacitor becomes a short-circuit and the voltage at its terminals will have to affect the others: a second failure can then predicted. And perhaps another one after that.

We can clearly see in these conditions a failure tree as the one illustrated in Figure 3.33.

In the case of parallel association, we have a current that is divided among several lines: if one of these uses up too much current, a thermal runaway phenomenon could take place[52] which will lower the impedance and/or the voltage at the terminals of the component thus used: the cause will then amplify the effect up to the destruction of the component: once more, two cases in point take place after the failure:

– either the component is short-circuited and all the parallel cells are nonfunctional[53];

– either the component is open and the current that used to circulate there must distribute in the surviving ones (that will not necessarily stay so...).

---

[52] This kind of phenomenon is equally at play in semi-conductors (typically in bipolar transistors).

[53] Loss of functionality for the supercapacitors and catastrophic failure for the batteries.

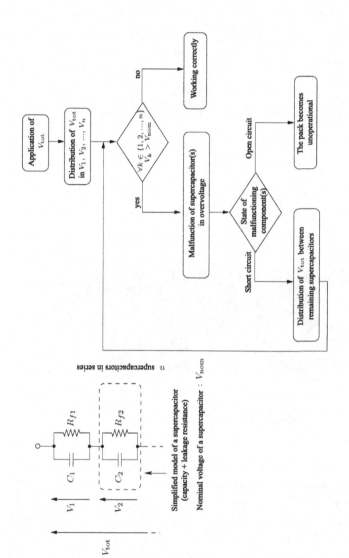

Figure 3.33. *Supercapacitors in series and associated failure trees (in case nominal voltages are surpassed)*

Diagram of a balancing circuit of two supercapacitors

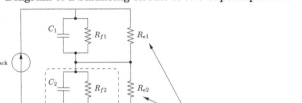

Balancing resistances
In practice, we will choose $R_{e1} = R_{e2} = R_e$
and we have $R_e \ll R_{f1}$ or R_{f2}

Supercapacitor
(capacity + failure resistance)

There are dispersions between the parameters of superconductors. Without the balancing
resistances, the V_{pack} voltage does not distribute equally between the components.

supercapacitors range of the pack element (source : Maxwell)

Figure 3.34. *Use of supercapacitors*

The main objective of a balancing circuit is thus to ensure a safe functioning of an association of components (in series or in parallel). This goal is achieved with the simplest solutions: passive circuits, which use resistances as shown in Figure 3.34. We note that this kind of assembly is very current when we wish to use supercapacitors. In fact, in medium and high power applications, the nominal voltages of unitary supercapacitors are too low in relation to the useful voltages of the application[54]. The supercapacitors packs commercialized by Maxwell are equipped with such circuits. We note that the presence of resistances leads to a discharge of supercapacitors even when none of the energy demands comes from the system. The simplicity of the passive balancing is thus obtained at the expense of energy efficiency.

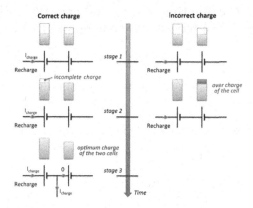

Figure 3.35. *Principle of active balancing of battery cells*

Another function of balancing circuits is to optimize the use of the capacities of each battery or the supercapacitor placed in a line (series association). This type of function can not be ensured by a purely passive circuit. The principle of such a balancing (that we will consider active) is illustrated in Figure 3.35. We can see there two battery cells placed in series and for each cell, a "bar graph" indicates not only its capacity but also its state of charge. This representation is then used to describe different states of the system:

– the cells are not fully charged (to the left as well as to the right);

54 The interface with a sur-voltage chopper or boost (see Chapter 1 of Volume 3 [PAT 15c]) will never make an energy transfer between a supercapacitor below 3 V and the continuous bus of an inverter under 200 V.

– to the left, the least capacitive cell is charged whereas the other is only partially charged. To the right, we continue the charge until the most capacitive cell is fully charged and in these conditions, there is the overcharge of the second cell;

– we continue the charge of the battery (to the left) but we disconnect the cell that is already fully charged so as to preserve only the one that we can fully charge with no risk of overcharging (as we have previously mentioned with the least capacitive cell).

Different circuit topologies of balancing circuits are possible. They are more or less complex:

– in terms of the number of power switches;

– in terms of electric connectors;

– in terms of control structure.

We can otherwise find balancing factors in which we discharge the overcharged cells in relation to others via a dedicated resistance and other active balancing factors in which the different elements can exchange energy (see the Renault license [GAG 14]).

3.5.3. *Battery chargers*

3.5.3.1. *General remarks*

The final indispensable element in the use of a battery is obviously the charger. This converter can be dedicated to this function or a reversible converter that is used to supply an equipment with the help of the battery when it functions in discharge mode: we can mention, for instance, the case of the inverter dedicated to the traction of an electric vehicle that can, when the phases of battery recharge, be used to ensure the functioning of the charger [LAC 13]. The converter associated to the battery is of x/DC type (DC/DC or AC/DC) with the battery connected "to a load". The topology of the converter can play a role in the nature of the quantities to control as we can ensure the supply of a battery following two modes:

– a current control (CC);

– a voltage control (CV).

We can use them either separately or successively and this depends on the technology of the battery and on the level of sophistication of the charger authorized by the application considered (compromise between cost/selling price, efficiency, researched lifespan, etc.).

Once the battery is charged (indicated potentially released by the BMS), we generally apply a voltage that allows us to provide a maintenance charge to the battery: this function is called *floating*. The level of the voltage depends on the technology of the battery but also on the temperature.

Generally speaking, the temperature always plays a role and it can be a critical one when we want to do a quick charge (risk of thermal runaway and fire hazard for Lithium-Ion batteries, water electrolysis and risk of explosion for lead batteries, premature aging of batteries "at best"). We classically qualify the charge depending on the capacity C (in A.h) of the battery; for example:

– a charge at $C/5$ will take five hours;

– a charge at $2C$ will take half hour.

Practical information (such as that presented in the following sections) for the recharge of batteries are regrouped at www.batteryuniversity.com.

3.5.3.2. *The case of lead batteries*

Nowadays, the charging of lead batteries is done in two subsequent steps (followed by a floating stage):

– a constant current phase (CC) during which we inject a current (generally limited to 5% of the battery's capacity): the voltage then increases gradually until it reaches a voltage between 2.35 and 2.40 V (per cel) for a temperature of 25°C;

– when the limit voltage of the previous stage is reached, it is maintained constant: the current then starts to decrease and stabilizes to a low value that no longer charges the battery but electrolyzes the water contained in the electrolyte (attention: risk of explosion);

– at the end of the charge (phase 2), the voltage is reduced between 2.25 and 2.28 V at 25°C (voltage lower than 2.34 V because over that, water electrolysis takes place) in order to preserve the charge: this phase is known as *floating*.

The voltage levels during stages 2 and 3 are modified from 0.005 V/°C (voltages that are increasing during the second stage and decrease during stage 3). We will note that it is recommended not to achieve a depth of discharge (DOD) higher than 80% with the technology of lead batteries to avoid sulfating, which is very rapid (several hours) for a strongly discharged battery. On the contrary, it is important not to overcharge the battery by remaining too long in phase 2 so as to avoid an important electrolysis contained in the battery (reduction of the level of the electrolyte, oxidation of parts non-submerged in electrodes; concentration of the acid and corrosion of electrodes, risk of explosion of the H_2/O_2 mixture).

We will note judging by the value of the current injected in phase 1 and the fact that the current is still lower in phase 2 that the charging time is long: 12 h to 16 h for example for salted batteries (typically for automobiles) and 36 to 48 h for batteries with a strong capacity for stationary applications.

3.5.3.3. *The case of Ni-MH batteries*

NiMH batteries can be recharged with (CC) current at levels of C/10 during 10 h to maximize their lifespan but the major difficulty is to detect the end of the charge (when the battery is charged at 100 %) in these conditions. In fact, the efficient methods for making this detection are:

– the observation of a voltage plateau;

– the observation of a voltage decrease;

– the observation of the battery heating.

These signs are only visible in the case of rapid charge (at $C/1$ and $4C$ respectively, that is for charges in 1 h to 15 min). Indeed in these cases, we see a significant decrease in the voltage at the terminals of the battery when it is past 100 % SOC. At the same time, the injection of a high current allows als to observe a more sensitive heating than at low current. In fact, the efficiency of the charge is close to 100 % for the phase during which the SOC is lower than 70 %. Beyond that, the battery starts to heat up progressively. The quick chargers are the most widespread in "widely available" applications, but it is worth noting that, even if they allow us to better manage the end of the charging, they can significantly reduce their lifespan. Certain charges use sequences during which the current gradually decreases with a beginning of a charge at $1C$ only to finish at the end of the charge with a maintenance current of no more than $C/20$.

3.5.3.4. *The case of Li-Ion batteries*

The charging of a Li-Ion battery is done in two phases, as for lead batteries[55]:

---

[55] However, for the latter, a CV type supply is only possible in "rudimentary" chargers.

– a first phase with constant curent (CC) ensuring a rapid charging at 80% (with a current of $C/2$ to $1C$, hence a phase during less than two hours towards less than one hour);

– a second phase with constant voltage during which the current decreases to achieve its value at th end of the charging.

The voltage at the end of the charging can vary between 4.1 and 4.2 V depending on the technology of the cell but tolerances are low (typically ± 50 mV), therefore requiring a precise controle of the part of the charger. We note that in practice, at the beginning of the charging, the voltage in the cell does not have to be lower than 3 V. Below this value, we find ourselves in a deep discharge situation that can destroy the element[56].

3.5.3.5. *Regarding the memory effect*

The memory effect is a well known phenomenon for Ni-Cd and Ni-MH technologies. In theory, it is possible that all the types of batteries are subjected to it but lead batteries and Li-Ion batteries are not sensitive to it either at all or they are only very little sensitive to it. There is a seemingly decrease in the capacity of the battery because of the partial and repeated discharges that change the structure of the reactives involved in the charge/discharge cycle in relation to those that are never used. It is consequently recommended to do a full discharge of Ni-Cd and Ni-MH batteries before recharging them so that they do not lose capacity.

In fact, the capacity is not really lost but the voltage delivered is lower once the non-cycled part is reacher. Physically, this memory effect translates as the increase in the size of crystals, reducing their exchange surface and

---

56 Here, the situation is analogous to that of lead batteries for which a deep discharge triggers a quick sulfating process (i.e. in several hours).

hence their reactivity. This memory effect has been emphasized the first time by NASA satellites that were recharged in a cyclical manner with photovoltaic panels with each "day/night" cycle of the satellite, whereas the discharge was only partial during the "night".

3.5.3.6. *Integrated circuits*

Integrated circuits are developed to pilot battery charges in all of the power technologies as we can see, for instance, on the website of Texas Instruments (see Figure 3.36). These circuits can sometimes integrate the power switches of the converter connected to the battery in order to decrease the number of required components and reducing the size of the global circuit: often, the main objective researched in "widely available" products of low power such as mobile phones, tables and other mobile devices supplied by a battery. In these cases, the use of the 5 V supply of a USB port of the computer is very frequent (not to say systematic) and several circuits offer this possibility.

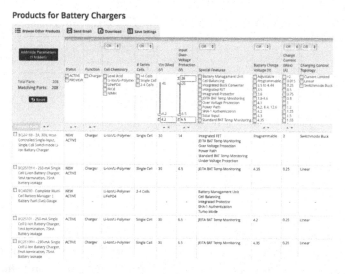

Figure 3.36. *Page dedicated to the selection of an integrated charger control circuit (Texas Instruments – www.ti.com)*

Let us note that the "battery charger" function is proposed by more manufacturers than the simple "Battery Fuel Gauge" (already mentioned in section 3.4). Here is a non-exhaustive list of the companies proposing such integrated circuits:

– Texas Instruments;

– Analog Devices;

– Maxim Integrated;

– linear Technology;

– ST Microelectronics;

– microchip;

– Fairchild Semiconductors;

– Intersil.

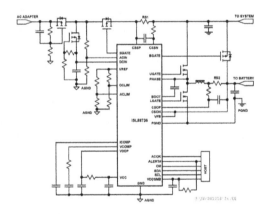

Figure 3.37. *Charger based on an Intersil ISL88736 circuit (Buck converter with synchronous recovery)*

If Texas Instruments is very well known in this field, there are certain other manufacturers that offer a complete component range. As this has been indicated in the section dedicated to monitoring, the company Intersil proposes not only functions relative to BMS (monitoring, among other things, the state of charge) but also disposes of solutions for the charging of batteries (either single cell or multiple cell batteries), especially for high power applications (for instance in the industry and the automobile sector). To conclude this chapter (and this volume) with remarks on an electronic power converter used as a charger, we can see in Figure 3.37 a charger based on an Intersil circuit ISL88736 where we recognize a *Buck* type chopper (with synchronous recovery)[57].

57 Allowing us to have a high efficiency with low voltage because by suppressing the voltage drop of the freewheel diode by putting in conduction the low transistor.

Appendix 1

Uncertainty Calculation

A1.1. Sources of errors and precision of measuring devices

A1.1.1. *Introduction*

The measures, whether they are done by measuring devices or by dedicated circuits in applications, are always riddled with errors. It is therefore worth considering them so as to evaluate how much we can trust the numeric values that we use in certain calculations. Indeed, even a low error can have a significant impact in sensitive calculations. The best example that we find in power electronics is the assessment of the efficiency of the converter. This seemingly simple problem is based on the classic definition of efficiency:

$$\eta = \frac{P_{\text{out}}}{P_{\text{in}}} \qquad \text{[A1.1]}$$

In fact, it is often very delicate because of the small difference between the two powers at play.

Indeed, as soon as the losses are reduced[1], we have $P_{\text{out}} \simeq P_{\text{in}}$. During measurements, we will generally obtain a value

---

1 Which is, basically, the main goal of our research.

associated to an uncertainty linked to the measuring device. To illustrate our point, let us consider the case where the input power is exactly 1 kW and the efficiency of the converter is 99 %, we can therefore evaluate the output power: it will be 990 W. These values are obviously impossible to measure exactly. In practice, we will consider here that we use precise wattmeters (with its calibration up to date) allowing us to measure the powers with a precision of 0.1 %. Admitting that the measurements are perfect (that is, unbiased) the two wattmeters placed in the input and output will respectively give us 1 kW and 990 W but we are not supposed to know if these are good results. We only know that the good result for each measurement is found in an interval of $\pm 0.1\%$ around the indicated value. In these conditions, we can provide the following frameworks for input and output powers:

– input, $999\,\text{W} \leq P_{\text{in}} \leq 1001\,\text{W}$;

– output, $989.01\,\text{W} \leq P_{\text{out}} \leq 990.99\,\text{W}$.

Once the two frameworks have been defined, the "optimistic" developer will be able to consider that the output power is the largest possible and the input power is the smallest to obtain the maximum efficiency $\eta_{\text{max}} = \frac{990.99}{999} \simeq 0.99198$ whereas the "pessimistic" one will take the maximum input power and the minimum output power to reach a minimum efficiency $\eta_{\text{min}} = \frac{989.01}{1001} \simeq 0.98802$. We therefore have a nominal efficiency of 99 % which carries, in fact, a relative uncertainty of the order 0.4 % ($\pm 0.2\%$). We then see that the uncertainty of this magnitude is the double of the uncertainty on the measures carried (powers). These uncertainties are not in fact proportional[2] as we will see now considering that the wattmeters are not more precise at 0.1 % but only at 1 %. We can then redo the same power framework calculations, which this time result in:

---

2 This is completely normal given that the relation [A1.1] is not linear.

– input $990\,\text{W} \le P_{\text{in}} \le 1010\,\text{W}$;

– output, $980.1\,\text{W} \le P_{\text{out}} \le 999.9\,\text{W}$.

We can then see that the minimum efficiency is equal to $\frac{980.1}{1010} \simeq 0.97$, which can seem satisfying since the good value is 0.99. However, we find ourselves in a significantly less physically credible situation if we calculate the maximum efficiency since the maximum output power is higher than the minimum input power: we then have $\eta_{\text{max}} = \frac{999.9}{990} \simeq 1.01$. The uncertainty this time is $\pm 2\,\%$, which is embarrassing because we find ourselves at $1\,\%$ loss in the converter.

Here, the uncertainty calculation allows us to determine if we will be capable (with a given level of measurement precision) to extract the information researched with a satisfying level of precision, in particular to avoid evident physical incoherences such as a efficiency higher than 1.

In the case of measuring devices, the precisions per measuring caliber are specified by the manufacturer. They guarantee them for a given period (usually one to two years) and propose calibrating services to mantain the performance throughout the life of these devices. In the case of specific circuits, we will have to examine ourselves the uncertainties starting from the tolerances of components, noises present in the circuits (thermal noise, shot noise, $1/f$ noise, etc.), time leeway (of all the components: resistances, voltage references, etc.) as well as leeways depending on the temperature. The objective of this annex is to introduce or recall physical notions about the uncertainty sources in an electric circuit and to present the mathematical tools allowing us to quantify the uncertainty of an "output" magnitude starting from the "input" quantities supposed to be known or characterized previously.

A1.1.2. *Uncertainties in electric circuits*

As we have already mentioned in Chapter 1, electronic circuits are subject to behavior uncertainties because of the components that make it up:

– these uncertainties can be observed from one electronic assembly to another, supposedly identical, one;

– these uncertainties can be observed on a unique assembly placed several times in the same trial conditions, while giving a different result each time.

These aspects have been introduced with the notion of faithfulness and fairness and it has been indicated that the first type of uncertainty could be "relatively slightly higher" thanks to a calibration at the end of the production to correct the unavoidable dispersions of each component (tolerance on the characteristics, in particular resistances).

In contrast, the second source of uncertainty is the fact of the data that has not been considered in the modeling (data that is often difficult to take into account) such as:

– temperature;

– degree of humidity;

– fluctuations in supply;

– noise in the circuits (resistances and semi-conductors);

– the past of the device.

In the precision circuits, we are careful to minimize the impact of this type of uncertainty but it is systematically the phenomenon that limits the performances of a measuring chain.

We will therefore distinguish two types of modelings for these uncertainties:

– bounded errors based models: the classic representation of the precision of a measuring device or the tolerance of a resistance;

– stochastic modelings: we often hypothesize a white Gaussian noise (but this is not always the case).

In the case of noise, the hypothesis of a Gaussian distribution is often justified because the noise is the macroscopic contribution of a large number of microscopic phenomena that are generally independent. In these conditions, the law of large numbers applies and the distribution of levels picked out from the magnitude considered (i.e. a power or a voltage) follows a normal law (so a Gaussian law) whose standard deviation σ we can easily assess via experiments. In contrast, the "whiteness" of the noise is often debatable: there is a hypothesis according to which the power of the signal is uniformly distributed on all the frequencies. In fact, the *Power Spectral Density* (PSD)[3] $\delta_{xx}(\tau)$ of a random signal $x(t)$:

$$\Gamma_x(\omega) = \mathcal{F}\left[\delta_{xx}(\tau)\right] = \int_{\mathbb{R}} \delta_{xx}(\tau) . e^{-j\omega\tau} . d\tau \qquad [A1.2]$$

where the self-correlation funciton $\delta_{xx}(\tau)$ is the mathematical expectation of the product of $x(t)$ times its offset conjugate of a duration τ:

$$\delta_{xx}(\tau) = \mathbb{E}\left[x(t).x^*(t-\tau)\right] = \lim_{T \to \infty} \int_{-T}^{+T} x(t).x^*(t-\tau).dt \qquad [A1.3]$$

In the case of real signals, the conjugation does not play any role, the mathematical expectation of the product

---

3 Do not confuse with *Digital Signal Processor* which is the exvoltage to random signals of the notion of Fourier transform for deterministic signals, is the Fourier transform of the self correlation function.

$x(t).x(t - \tau)$ simply evaluates the resemblance of the signal to itself when it is offset by a given duration τ. In the case of white noise, $X(\omega)$ is a constant because $\delta_{xx}(\tau)$ is a Dirac impulsion (whose weight is simply the variation of $V = \sigma^2$) hence a "function" (or rather a *distribution*) that is null everywhere except in zero: this means that the signal is no longer similar to itself as soon as we introduce an offset $\tau \neq 0$. This theoretical noise (which, however, models well enough thermal noise, for instance), has the advantage of being easy to manipulate in calculations. The "normality" property (Gaussian noise) is still the most important as it does not vary by linear filtering: consequently, if we apply a Gaussian noise at the entry of a linear filter, the exit of this filter will also be a Gaussian noise. Let us remember here the formula which allows us to establish a link between the PSDs of signals in input and output (respectively $\Gamma_x(\omega)$ and $\Gamma_y(\omega)$) of a transfer function filter $H(\omega)$:

$$\Gamma_y(\omega) = |H(\omega)|^2 . \Gamma_x(\omega) \tag{A1.4}$$

Having gone through a low-pass filter for instance, a Gaussian white noise remains Gaussian but it is no longer white: we then speak of *pink noise*.

A1.1.3. *Acquisition uncertainties*

Acquisitions are also subject to uncertainties. In the case of old measuring devices, the reading on a dial switch introduced a human error that could be more or less big:

– parallax error because of the misalignment with the needle (corrected by aligning the needle with its reflection in the mirror included in the dial indicator);

– reading with both eyes open (the highest precision is obtained with one eye shut and the other perfectly aligned to the needle and its reflection);

– approximation of the position of the needle between two graduations.

Even if this technology can seem obsolete, the latter point is always power in modern digital devices. Indeed, the analog signals are generally processed by a digital analog conversion stage (ADC for *Analog Digital Converter*). The principle of such converters is to replace the analog voltages (an interval of real numbers) in numbers on a given number of bits (so in finite quantity). Consequently, a number does not represent a voltage but the interval of voltages that we call *quantum*. Generally speaking, an ADC disposes of an input range limited between a voltage V_{min} and a voltage V_{max}. We then note the difference $\Delta V = V_{max} - V_{min}$ and this interval is broken down in sub-intervals of a size $q = \frac{\Delta V}{2^N}$ (the *quantum*) where N is the number of bits of the converter. We can then draw up a list of power values N and note the number of codes $N_c = 2^N$ available:

– for $N = 8$, we have $N_c = 256$;

– for $N = 10$, we have $N_c = 1\,024$;

– for $N = 12$, we have $N_c = 4\,096$;

– for $N = 14$, we have $N_c = 16\,384$;

– for $N = 16$, we have $N_c = 65\,536$;

– for $N = 20$, we have $N_c = 1\,048\,576$;

– for $N = 24$, we have $N_c = 16\,777\,216$;

– for $N = 31$, we have $N_c = 2\,147\,483\,648$[4].

---

[4] As far as we know, the highest resolution on an integrated monolithic ADC (Texas Instruments ADS1282).

As this has also been mentioned, generally speaking, the highest frequencies of sampling can only be achieved by ADC with a limited resolution (generally 8 bits in Flash technology, in digital oscilloscopes for instance). In contrast, the higher resolutions use either slow (multiple) technologies, or adaptable technologies ($\Sigma - \Delta$ with the possibility of modifying the number of bits).

In all the cases in the figure, it is worth noting that the jump from a code to another corresponds exactly to the graduations of a dial indicator instrument between two graduations the real level of the signal is more or less unknown. In fact, everything takes place as if we added noise to the real signal, thus forcing the acquired signal to evolve "in the steps of a stair". The maximum deviation between the ideal signal and the serrated signal being equal to $q/2$, we can evaluate this noise by its root mean square (RMS) value B_{RMS} (or in the context of random deviation-type signals). Different calculation methods exist to reach the result but here we will only mention this:

$$B_{RMS} = \frac{q}{\sqrt{12}} = \frac{\Delta V}{2^{N+1}\sqrt{3}} \qquad \text{[A1.5]}$$

If we now place ourselves in the most favorable situation for analyzing a (periodic) variable signal: using the full scale to represent it (peak-to-peak amplitude equal to ΔV), we can evaluate its efficient value S_{RMS} and compare it to that of the noise. We will consider here the case of a sinusoidal signal whose amplitude is therefore equal to $\frac{\Delta V}{2}$ and consequently with an efficient value equal to $\frac{\Delta V}{2\sqrt{2}}$. In these conditions, the signal over noise ratio (SNR for *Signal/Noise Ratio*) has the value:

$$\text{SNR} = \frac{\frac{\Delta V}{2\sqrt{2}}}{\frac{\Delta V}{2^{N+1}\sqrt{3}}} = \frac{2^{N+1}\sqrt{3}}{2\sqrt{2}} = 2^N \sqrt{\frac{3}{2}} \qquad \text{[A1.6]}$$

It is then more classic to express this ration in dB by noting $\text{SNR}_{\text{dB}} = 20 \log (\text{SNR})$ and therefore:

$$\text{SNR}_{\text{dB}} = 20.N.\log 2 + 20.\log \sqrt{\frac{3}{2}} \simeq 6,021.N + 1,761 \,[\text{A1.7}]$$

We will again be able to draw the list of SNR_{dB} (in the given dB) for the power N values:

– for $N = 8$, we have $\text{SNR}_{\text{dB}} \simeq 50\,\text{dB}$,

– for $N = 10$, we have $\text{SNR}_{\text{dB}} \simeq 62\,\text{dB}$,

– for $N = 12$, we have $\text{SNR}_{\text{dB}} \simeq 74\,\text{dB}$,

– for $N = 14$, we have $\text{SNR}_{\text{dB}} \simeq 86\,\text{dB}$,

– for $N = 16$, we have $\text{SNR}_{\text{dB}} \simeq 98\,\text{dB}$,

– for $N = 20$, we have $\text{SNR}_{\text{dB}} \simeq 122\,\text{dB}$,

– for $N = 24$, we have $\text{SNR}_{\text{dB}} \simeq 146\,\text{dB}$,

– for $N = 31$, we have $\text{SNR}_{\text{dB}} \simeq 188\,\text{dB}$.

We will note that these are theoretical values and they are not generally obtained in practical situations: it is often rare to exploit an ADC in its full scale. On the contrary, we will try to optimize its use by having as many signals as possible between the second half of the full scale to exploit strong weight bit. Evidently, it is not always the case but it would be reasonable to adapt the measuring caliber(s) so that the measuring range is exploited to the maximum for the most frequent situations (for instance for the nominal voltage and/or the nominal power at the output of a breakdown supply).

Finally, it is worth noting that for high resolutions (20 bits or more) the quanta are in practice extremely narrow and consequently, the noise contained in the analog signal plays a very significant role. Even if the converters that have such resolutions are slow, we will also have to reduce the bandwidth so as to minimize the amplitude of the noise because if we do not, the least significant bits will essentially be useless random... sequences of 0 and 1. And if this is not done in the analog field, we will have to do such a filtering in the digital field in order to be able to exploit the data ("high resolution" mode of digital devices).

REMARK A1.1.– We will not go back on the time sampling issue here when we use an analog converter, given that this aspect of the problem (i.e. aliasing and Shannon theorem) was presented in the appendix "Elements of spectral analysis" in Volumes 2 and 4. However, it is evident that this phenomenon cannot introduce biases in measurements if it is not correctly considered in the numerical processing chain of the signal.

A1.1.4. *Uncertainty representation, vocabulary and conventions*

A1.1.4.1. *Theoretical analysis of uncertainties and probabilities*

In the case of an uncertainty caused by a large number of random phenomena, the distribution of the measure follows a normal law that can be approached experimentally by a number of measures high enough to establish its standard deviation and thus wholly characterize the uncertainty. This is because a normal law is completely described by the average and by the standard deviation.

In the case of measuring devices for which we have a strict framework of the measurement by an interval, the notion of normal law disappears and we can consider that the distribution of the uncertainty is uniform in the interval considered. If the interval is somewhere between a value x_{min}

and a value x_{max}, we then have a probability density $p(X)$ of obtaining the measure having the value X expressed thus:

$$p(X) = \begin{cases} 0 \text{ if } X < x_{min} \\ \frac{1}{x_{max}-x_{min}} \text{ if } x_{min} \leq X \leq x_{max} \\ 0 \text{ if } X > x_{max} \end{cases} \qquad [\text{A1.8}]$$

We can easily verify that the integer $p(x)$ on all the possible values of X (i.e. the real axis) is equal to 1. Then, we can verify that the probability of having a value lower than x_{min} or higher than x_{max} is null because, by definition, X can not leave the interval $[x_{min}; x_{max}]$. On the contrary, in this interval, all of the values are equi-probable. We can then calculate the moments of the order 1 and 2 of his uniform distribution:

$$\mathbb{E}[X] = \int_{\mathbb{R}} X.p(X).dX = \frac{x_{min} + x_{max}}{2} = \overline{X} \qquad [\text{A1.9}]$$

In fact, the moment of order 2 is a *centered moment* (for which we replace x by a random centered variable $X - \overline{X}$, that is with a null average value). We thus define the variance V_X of the random variable X:

$$V_X = \mathbb{E}\left[\left(X - \overline{X}\right)^2\right] = \int_{\mathbb{R}} \left(X - \overline{X}\right)^2.p(X).dX$$

$$= \frac{\left(x_{max} - x_{min}\right)^2}{24} \qquad [\text{A1.10}]$$

And its square root is called standard deviation σ_X:

$$\sigma_X = \frac{x_{max} - x_{min}}{\sqrt{12}} \qquad [\text{A1.11}]$$

In fact, we find the efficient value of the quantification noise that has been presented in section A1.1.3 (see the equation [A1.5]) since $x_{max} - x_{min}$ is perfectly equivalent to the quantum of a digital-analog converter. In a statistic framework, the uncertainty can be described with the help of a standard deviation (we speak of *standard-uncertainty*) but it is clear that the framing of the average value $\bar{X}$ of the measures x_i in an interval $[\bar{X} - \sigma_X; \bar{X} + \sigma_X]$ is far from covering the entire set of measures. Indeed, in the case of a normal law, we can only find 68.27% of the measurements. In order to obtain a higher confidence rate T_c, we widen this interval: we then speak of *widened uncertainty*, characterized by a widening coefficient k. We can thus give the percentage measurements included in the interval $[\bar{X} - k\sigma_X; \bar{X} + k\sigma_X]$ for different values of k (always by a normal law):

– for $k = 2$, we have 95.45% of measurements in the interval;

– for $k = 3$, we have 99.73% of measurements in the interval.

OBSERVATION A1.1.– In practice, we generally look for a confidence rate T_c of 95%. A widening coefficient lower than 2 seems therefore to suffice (1.96 more precisely). Nevertheless, the norm predicted by the AFNOR (Agence Franchaise de NORmalisation) consists of taking a widening coefficient of 2.

These results can obviously be found starting from the *distribution function*[5] $\Phi(x)$ defined as follows:

$$\Phi(x) = \int_{-\infty}^{x} \frac{1}{\sqrt{2\pi}} e^{-\frac{t^2}{2}} dt = \mathbb{P}(X < x) \qquad \text{[A1.12]}$$

This function allows us to calculate (for a normal standard deviation distribution equal to 1) the probability that the random variable X is loewr than a given value x. This probability noted $\mathbb{P}(X < x)$ (or also Prob($X < x$)) is,

_____

5 Associated to the famous Gauss error function $\mathrm{erf}(x)$ (*"error function"*).

evidently, as any probability a value comprised between 0 and 1 and the function $\Phi(x)$ is a monotonous increasing function. From this, we can then deduce that:

$$\mathbb{P}\left(x_a < X < x_b\right) = \mathbb{P}\left(X < x_b\right) - \mathbb{P}\left(X < x_a\right)$$

$$= \Phi\left(x_b\right) - \Phi\left(x_a\right) \qquad \text{[A1.13]}$$

The distribution function being associated to the reduced law (i.e. of a unitary standard-deviation), it simply corresponds to a real centered variable Y divided by its standard deviation $\sigma = \sigma_Y$ ($X = \frac{Y}{\sigma_Y}$).

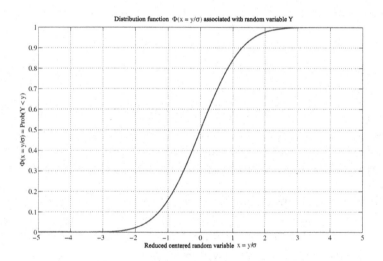

Figure A1.1. *Distribution function $\Phi(x)$ – Centered reduced normal distribution $\mathcal{N}(\overline{X} = 1, \sigma_X = \sigma = 1)$*

A1.1.4.2. *Experiments and statistics*

The experimental assessment of a distribution requires several acquisitions to hope to converge towards the probability law that is being researched. Several processes where a great number of parameters are at play presenting a random variation naturally (or even mathematically) tend towards a normal law. This presents the advantage of being characterizable by only two parameters: average and standard deviation. Unfortunately, these two parameters are inaccessible in practice because, to obtain a convergence of experimental values towards asymptotic values, we would need an infinity of experimental data. The truncation (at N experiments) of the sum of samples to obtain the *experimental average*:

$$\overline{x}_{\exp} = \frac{1}{N} \sum_{k=1}^{n} x_k \qquad\qquad [A1.14]$$

as well as the one used for the *experimental standard deviation*:

$$\sigma_{\exp} = \sqrt{\frac{1}{N-1} \sum_{k=1}^{N} (x_k - \overline{x}_{\exp})^2} \qquad\qquad [A1.15]$$

inevitably introduce biases. In the equation [A1.15] we will note the division by $N-1$ that is the norm for the definition of standard deviation whereas if we carry out a division by N, we rather speak of *average quadratic deviation*.

It is however possible to correct the results established in the case of a low number N of sampling for extrapolating the normal underlying law. In fact, we can give a framework of the theoretical average (which we note $\overline{x}$) starting from averages and experimental standard deviations ($\overline{x}_{\exp}$ and $\sigma_{\exp}$ respectively) and a factor t called student coefficient that we can see (in Table A1.1) that it depends on the number of N samples available and of the desired confidence rate T_c for the framing:

$$\overline{x} = \overline{x}_{\exp} \pm \frac{t.\sigma_{\exp}}{\sqrt{n}} \qquad\qquad [A1.16]$$

Student Coefficients t	Confidence rate T_c								
	50	80	90	95	98	99	99,5	99,8	99,9
1	1,00	3,08	6,31	12,7	31,8	63,7	127	318	637
2	0,82	1,89	2,92	4,30	6,96	9,92	14,1	22,3	31,6
3	0,76	1,64	2,35	3,18	4,54	5,84	7,45	10,2	12,9
4	0,74	1,53	2,13	2,78	3,75	4,60	5,60	7,17	8,61
5	0,73	1,48	2,02	2,57	3,36	4,03	4,77	5,89	6,87
6	0,72	1,44	1,94	2,45	3,14	3,71	4,32	5,21	5,96
7	0,71	1,41	1,89	2,36	3,00	3,50	4,03	4,79	5,41
8	0,71	1,40	1,86	2,31	2,90	3,36	3,83	4,50	5,04
9	0,70	1,38	1,83	2,26	2,82	3,25	3,69	4,30	4,78
10	0,70	1,37	1,81	2,23	2,76	3,17	3,58	4,14	4,59
11	0,70	1,36	1,80	2,20	2,72	3,11	3,50	4,02	4,44
12	0,70	1,36	1,78	2,18	2,68	3,05	3,43	3,93	4,32
13	0,69	1,35	1,77	2,16	2,65	3,01	3,37	3,85	4,22
14	0,69	1,35	1,76	2,14	2,62	2,98	3,33	3,79	4,14
15	0,69	1,34	1,75	2,13	2,60	2,95	3,29	3,73	4,07
16	0,69	1,34	1,75	2,12	2,58	2,92	3,25	3,69	4,01
17	0,69	1,33	1,74	2,11	2,57	2,90	3,22	3,65	3,97
18	0,69	1,33	1,73	2,10	2,55	2,88	3,20	3,61	3,92
19	0,69	1,33	1,73	2,09	2,54	2,86	3,17	3,58	3,88
20	0,69	1,33	1,72	2,09	2,53	2,85	3,15	3,55	3,85
22	0,69	1,32	1,72	2,07	2,51	2,82	3,12	3,50	3,79
24	0,68	1,32	1,71	2,06	2,49	2,80	3,09	3,47	3,75
26	0,68	1,31	1,71	2,06	2,48	2,78	3,07	3,43	3,71
28	0,68	1,31	1,70	2,05	2,47	2,76	3,05	3,41	3,67
30	0,68	1,31	1,70	2,04	2,46	2,75	3,03	3,39	3,65
40	0,68	1,30	1,68	2,02	2,42	2,70	2,97	3,31	3,55
50	0,68	1,30	1,68	2,01	2,40	2,68	2,94	3,26	3,50
60	0,68	1,30	1,67	2,00	2,39	2,66	2,91	3,23	3,46
70	0,68	1,29	1,67	1,99	2,38	2,65	2,90	3,21	3,44
80	0,68	1,29	1,66	1,99	2,37	2,64	2,89	3,20	3,42
90	0,68	1,29	1,66	1,99	2,37	2,63	2,88	3,18	3,40
100	0,68	1,29	1,66	1,98	2,36	2,63	2,87	3,17	3,39
200	0,68	1,29	1,65	1,97	2,35	2,60	2,84	3,13	3,34
300	0,68	1,28	1,65	1,97	2,34	2,59	2,83	3,12	3,32
500	0,67	1,28	1,65	1,96	2,33	2,59	2,82	3,11	3,31
1000	0,67	1,28	1,65	1,96	2,33	2,58	2,81	3,10	3,30
∞	0,67	1,28	1,64	1,96	2,33	2,58	2,81	3,09	3,29

The leftmost column is labeled "Degrees of freedom $(N-1)$".

Table A1.1. *Student coefficients*

We then find a framework analogous to that of tolerances for the components (resistances, conductors, etc.) or on the precision of the measures provided by the metrology devices. The term $\Delta x = \frac{t.\sigma_{exp}}{\sqrt{n}}$ is then called *absolute uncertainty* and the ratio $\Delta x / \bar{x}$ relative uncertainty (often noted in % or, for very low values, in ppm, or ppb, respectively[6].

A1.1.4.3. *Notations for values and uncertainties*

When a magnitude is given with an uncertainty, common sense tells us not to give a value whose significant numbers are definitely below the uncertainty threshold. It is indeed useless to give ten numbers after the comma when we say that the uncertainty impacts the third one, for instance. There is, in fact, a conventions tht consists of taking as significant numbers all of those that are not affected by uncertainty and add one. We then complete with the uncertainty of the form $\pm xx \times 10^y$ where "xx" represents two significant numbers. As an example, here is an uncertain value put in a correct form:

$$\underbrace{3.141592}_{\text{certain}} \quad \underbrace{7}_{\text{uncertain}} \quad \underbrace{\pm 0.000015}_{\text{uncertainty}} \tag{A1.17}$$

We can see in this example that the uncertainty has only two significant numbers and that the last number of the value (7) is potentially modified by the uncertainty.

OBSERVATION A1.2.– We round up the values in the usual way (a sequence of numbers "x5..." will be rounded up with the digit "x+1"). On the other hand, the uncertainty is (more prudently) rounded up systematically to the superior integer: for instance, the sequence "xyz" having to be rounded up to the digit x will always be rounded up to $x+1$ (even if "yz" has the value "01" for example).

---

6 ppm means "parts per million" whereas ppb means "parts per billion".

A1.2. Uncertainties of composed quantities

When we dispose of uncertain values for quantities that come up in a mathematical expression whose precision we wish to evaluate, we can apply set methods or more simply, evaluate the worst case scenario as we did in Chapter 1 for the evaluation of the CMRR (Common Mode Rejection Rate) of an subtracting assembly with an operational amplifier for which we have considered that the R_i resistances of the assembly were all of the form $R_0 + \Delta R_i$. This approach allows us to obtain a framework of the value with a high confidence rate[7] but this "handcrafted" approach (from situation to situation) of the evaluation of uncertainties can be difficult to use for "handmade" calculations. Indeed, in the case of more complex systems, determining the combination that produces the worst scenario can prove to be a difficult task. Obviously, using software, a "rough" analysis can overcome this difficulty. Despite everything, the use of other methods (potentially approximate) can make the analysis more simple and even feasible without any software tools.

These calculations are based on the principle that uncertain (if not even random) variables are independent. This is a reasonable hypothesis when we are interested in the properties of strictly distinct components but this is not always the case in electronics: the uncertainties can be produced by deviations due to the temperature and we can assume that the room temperature in which different components are found will affect them in the same manner: in fact we will rather speak of correlation. It is therefore worth defining the two notions of independence and correlation.

---

7 We could say we were certain that the tolerances given for the resistances by the constructors give a framework of all the values possible with a confidence rate of 100 %. We will always be able to inquire about the notion of absolute certainty.

A1.2.1. *Independence and correlation*

The notion of independence is a notion that we can intuitively define as the property of two random variables that have no connection. When we speak of two independent events A and B and their respective probabilities of occurrences $\mathbb{P}(A)$ and $\mathbb{P}(B)$, the definition allows us to describe that the occurrence probability of the event "A and B" (noted $A \cap B$). The independence of two variables is difficult to prove in practice: there is no truly reliable index that proves this property when the correlation is an experimentally measurable magnitude (at least in a close way). There is a "normalized covariance" $\mathrm{Cor}(X, Y)$ in the sense that we take the equation [A1.10] and adapt it for the two random variables X and Y (instead of only one) to calculate the covariance $\mathrm{Cov}(X, Y)$:

$$\mathrm{Cov}(X, Y) = \mathbb{E}\left[(X - \overline{X}).(Y - \overline{Y})\right] = \mathbb{E}[X.Y] - \overline{X}.\overline{Y} \quad \text{[A1.18]}$$

OBSERVATION A1.3.– When the two variables X and Y are independent, we can note the following equivalence:

$$\mathbb{E}[X.Y] = \mathbb{E}[X].\mathbb{E}[Y] = \overline{X}.\overline{Y} \quad \text{[A1.19]}$$

and consequently, the covariance of these two variables is null.

Starting from the covariance, we obtain the correlation by dividing this result to the standard deviations of X and Y, respectively noted σ_X and σ_Y:

$$\mathrm{Cor}(X, Y) = \frac{\mathrm{Cov}(X, Y)}{\sigma_X.\sigma_Y} \quad \text{[A1.20]}$$

with:

$$\sigma_X = \sqrt{\mathbb{E}\left[(X - \bar{X})^2\right]} = \sqrt{\int_{\mathbb{R}} (X - \bar{X})^2.p(X).dX} \quad \text{[A1.21]}$$

and similarly:

$$\sigma_Y = \sqrt{\mathbb{E}\left[(Y - \bar{Y})^2\right]} = \sqrt{\int_{\mathbb{R}} (Y - \bar{Y})^2 . p(Y).dY} \quad [A1.22]$$

When the variables X and Y are scalar variables, the correlation $\mathrm{Cor}(X, Y)$ is a number comprised between -1 and 1 enabling us to determine if the variables are more or less inter-connected. If the number is close to zero, we say that the variables X and Y uncorrelated whereas when $\mathrm{Cor}(X, Y)$ tends towards 1 in absolute value (so therefore -1 or 1), we say that the variables are strongly correlated. We must, however, be prudent in our interpretation of this magnitude. When there is an affinity between the two quantities,the analysis of the correlation clearly allows us to identify a deterministic link between the two variables X and Y. Unfortunately, this is not a general result: we can not conclude, therefore, a null correlation, that there is no deterministic relation between two random variables as we can perfectly propose simple nonlinear relations that allow us to theoretically reach a null correlation.

In fact, thanks to this tool, we can qualitatively emphasize that when X is positive (or negative, respectively), Y is generally positive for a positive correlation or generally negative (or negative) for a negative correlation. In contrast, if we define Y based on X using a (deterministic) pair function, we wil reach a null correlation coefficient whereas the quantities are celarly coupled when we represent them by a "cloud of dots" with a certain number of "random" couple realizations (X, Y) as shown in Figure A1.2.

OBSERVATION A1.4.– It is evident that when two random variables X and Y are strongly correlated with a constellation of perfectly aligned dots (as the clouds in the extremes of the two lines in Figure A1.2), we can bring the representation of uncertainties to a single random variable instead of two.

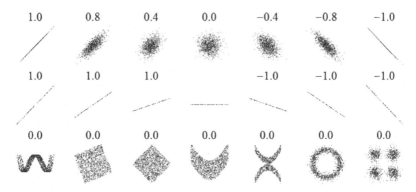

Figure A1.2. *Correlation coefficients for different types of couplings X, Y (source: Wikipedia)*

Considering the couplings between two random variables is the first stage in the evaluation of uncertainties on composed quantities that puts into play several uncertain variables. We then dispose of methods that allow us to calculate the uncertainty of the expression depending on the uncertainties of each variable, starting from an analytical expression of several variables. We will see in the following two sections that there are two approaches enabling us to carry out this type of study:

– a close method based on the calculation of partial deviations;

– the use of usual formulas.

A1.2.2. *Application of the partial deviations method*

A1.2.2.1. *Principles*

The application of the partial deviations method consists of establishing a development limited by a function with n

variables while supposing that the uncertainties are small variations applied to variables whose nominal values are knwon. This method therefore consists of carrying out a limited development (at the 1st order). We linearize the evolution of the function $f(\mathbf{x})$ around a point in a space of size n (noting that $\mathbf{x} = (x_1, x_2, \ldots, x_n)^t$):

$$df = \frac{\partial f}{\partial x_1} dx_1 + \frac{\partial f}{\partial x_2} dx_2 + \cdots + \frac{\partial f}{\partial x_n} dx_n \qquad [A1.23]$$

In fact, this equation allows us to determine (in the manner of a limited development) *hyperplane tangents* to the variety defined by $f(\mathbf{x})$. We can thus do a standard deviation calculation on the variations of f (note σ_f) around a nominal point $\mathbf{x}_0 = (x_{10}, x_{20}, \ldots, x_{n0})^t$ by calculating the partial deviations of f for each component in this point $\mathbf{x}_0$ and by making appear in this equation standard deviations σ_k afffecting each variable x_k (for $k \in \{1, 2, \ldots, n\}$):

$$\sigma_f^2 = \sum_{k=1}^{n} \left(\left. \frac{\partial f}{\partial x_k} \right|_{\mathbf{x}=\mathbf{x}_0} . \sigma_k \right)^2 \qquad [A1.24]$$

Similarly, the uncertainty Δf can be calculated by using the same formula by replacing the standard deviations σ_k with the corresponding Δ_k uncertainties (and of course by replacing σ_f with Δf). In fact, this method only functions if the (uncertain) variables considered are independent.

A1.2.2.2. *Example*

For illustrative purposes, let us take the case of a function f with four variables (x_1, x_2, x_3, x_4) whose definition is:

$$f(x_1, x_2, x_3, x_4) = x_1 \cdot \frac{x_2 - x_3}{x_4} \qquad [A1.25]$$

By applying the formula [A1.24] applied to the uncertainties Δf for the function and Δ_k for the variables x_k ($1 \leq k \leq 4$), we obtain:

$$\Delta f^2 = \sum_{k=1}^{4} \left(\left. \frac{\partial f}{\partial x_k} \right|_{\mathbf{x}=\mathbf{x_0}} \cdot \Delta_k \right)^2 \qquad \text{[A1.26]}$$

and it is therefore enough to evaluate the four partial deviations (considering that $\mathbf{x_0} = (x_{10}, x_{20}, x_{30}, x_{40})^t$):

$$\begin{aligned}
\left. \frac{\partial f}{\partial x_1} \right|_{\mathbf{x}=\mathbf{x_0}} &= \frac{x_{20}-x_{30}}{x_{40}} \\[4pt]
\left. \frac{\partial f}{\partial x_2} \right|_{\mathbf{x}=\mathbf{x_0}} &= \frac{x_{10}}{x_{40}} \\[4pt]
\left. \frac{\partial f}{\partial x_3} \right|_{\mathbf{x}=\mathbf{x_0}} &= -\frac{x_{10}}{x_{40}} \\[4pt]
\left. \frac{\partial f}{\partial x_4} \right|_{\mathbf{x}=\mathbf{x_0}} &= -x_{10} \cdot \frac{x_{20}-x_{30}}{x_{40}^2}
\end{aligned} \qquad \text{[A1.27]}$$

This method has the advantage of being very general and allows us to reach the following result:

$$\Delta f^2 = \left(\frac{x_{20}-x_{30}}{x_{40}} \right)^2 \cdot \Delta_1^2 + \left(\frac{x_{10}}{x_{40}} \right)^2 \cdot (\Delta_2^2 + \Delta_3^2)$$

$$+ \left(x_{10} \cdot \frac{x_{20}-x_{30}}{x_{40}^2} \right)^2 \cdot \Delta_4^2 \qquad \text{[A1.28]}$$

OBSERVATION A1.4.– We can easily see in this example that the uncertainties of a sum are not added. With this method, the square of the uncertainty is equal to the square sum of uncertainties in this case. We see that the degree of confidence of the interval obtained will be lower than the rate associated to the uncertainties of the components because we would naturally have the tendency to sum the uncertainties to place ourselves in the worst case (always true, particularly *when the uncertainties are correlated*). In fact, in the special case where all the uncertainties have the same amplitude Δx and in the case where we could do a sum of a high number of components n, the uncertainty of the sum would be of the form $\Delta x.\sqrt{n}$ which is a known result for the summing up of *independent random variables*. The prudent method would like for the uncertainties to be directly summed up but this approach would lead to largely overrated results for a large number of variables that are truly independent. The passing from a low number to a high number is therefore worth analyzing in order to adapt the approach to the desired degree of confidence.

A1.2.3. *Usual formulas*

General formulas can be established for simple mathematical expressions that often reappear in uncertainty calculations: in the occurrence, there are four arrhythmic operations (addition, subtraction, multiplication, division) but also the function "power". These useful results are regrouped in the Table A1.2 where we can see the expression of the composed magnitude X, its absolute uncertainty ΔX and its relative uncertainty $\frac{\Delta X}{X}$.

Operation	Composed magnitude X	Absolute uncertainty ΔX	Relative uncertainty $\frac{\Delta X}{X}$
Addition	$x_1 + x_2$	$\Delta x_1 + \Delta x_2$	$\frac{\Delta x_1 + \Delta x_2}{x_1 + x_2}$
Subtraction	$x_1 - x_2$	$\Delta x_1 + \Delta x_2$	$\frac{\Delta x_1 + \Delta x_2}{x_1 - x_2}$
Multiplication	$x_1.x_2$	$x_1 \Delta x_2 + x_2 \Delta x_1$	$\frac{x_1 \Delta x_2 + x_2 \Delta x_1}{x_1.x_2} = \frac{\Delta x_1}{x_1} + \frac{\Delta x_2}{x_2}$
Division	$\frac{x_1}{x_2}$	$\frac{x_1 \Delta x_2 + x_2 \Delta x_1}{x_2^2}$	$\frac{\Delta x_1}{x_1} + \frac{\Delta x_2}{x_2}$
Power	x_1^n	$n.x_1^{n-1}\Delta x_1$	$\frac{n.\Delta x_1}{x_1}$

Table A1.2. *Usual formulas for calculating uncertainties*

A1.3. Return to the calculation of efficiency

We will finish this appendix by returning to the example of the calculation of efficiency. However, this time, we will not limit ourselves to a single uncertainty calculation on efficiency based on input power measurements P_e and output measurements P_s of the device studied (here a continuous power machine, or CCM). In fact, we will consider that the input power is written:

$$P_e = U_{dc}.I_{dc} \qquad\qquad [A1.29]$$

where U_{dc} and I_{dc} are voltages applied to the induced terminals of the machine (in V and A respectively). As far as the output is concerned, the mechanical power provided is written:

$$P_s = C_m.\Omega_m \qquad\qquad [A1.30]$$

where C_m is the couple released (in N.m) and Ω_m is the rotation speed of the rotor (in rad/s). In the case of a machine with unitary efficiency, we can write the two following relations:

$$\begin{cases} U_{dc} = E = k.\Omega \\ C_m = k.I_{dc} \end{cases} \qquad\qquad [A1.31]$$

where k is the electromechanic conversion coefficient of the machine (expressed in V/rad/s or in N.m/A). Evidently, this is not the case for a real machine for which there are losses:

– in copper (because of the electric resistance of the wires);

– in the magnetic parts (we speak of "iron" losses of two types: hysteresis and Foucault powers[8]);

– the mechanical losses (because of the friction between the collector and the brush) but also in rolling motions, because of aerodynamic friction.

---

8 These two types of losses are particularly distinct in their variation depending on the frequency (respectively in f and in f^2 upon first analysis).

By admitting now that we uniquely dispose of separate devices for measuring the input powers (a voltmeter and an anmeter) and output (meter-couple and tachometer), we propose to determine the uncertainty of the efficiency $\eta = P_s/P_e$ of the machine on the basis of the precisions of different devices.

For numerical applications, we consider here that the precision of the measurements for voltage U_{dc} and the power I_{dc} are both $\pm 0.01\,\%$ whereas the precisions on the rotation speed and on the couple are only $\pm 0.1\,\%$. As for the nominal values measured, we consider that:

– the voltage U_{dc} is $200\,$V;

– the power is $50\,$A;

– the couple is $45\,$N.m;

– the rotation speed is $212\,$rad/s.

We can easily deduce that the input power is evaulated at $10\,$kW whereas the output power is estimated at $9\,540\,$W. We can then deduce a efficiency of $95.4\,\%$ so all we need to define is precision.

A1.3.1. *The intuitive "approach"*

This case is simple enough to be able to apply an intuitive approach: the highest efficiency is obtained for the highest output power and for the smallest input power, whereas the situation is inversed if we seek the smallest efficiency. On the basis of this starting hypothesis, we are led to calculate the "min" and "max" powers in both input and output:

$$
\begin{cases}
P_{e\min} = U_{dc\min}.I_{dc\min} \\
P_{e\max} = U_{dc\max}.I_{dc\max} \\
P_{s\min} = C_{m\min}\Omega_{m\min} \\
P_{s\max} = C_{m\max}\Omega_{m\max}
\end{cases}
\qquad [\text{A1.32}]
$$

Evidently, these "min" and "max" quantities are deduced from values obtained with the measuring devices on the basis of their precision. Generally speaking, we will give a minimal (or maximal) magnitude in the form:

$$\begin{cases} x_{\min} = x_0 - \Delta x = x_0 \left(1 - \frac{\Delta x}{x_0}\right) \\ x_{\max} = x_0 + \Delta x = x_0 \left(1 + \frac{\Delta x}{x_0}\right) \end{cases} \quad \text{[A1.33]}$$

where Δx is the relative uncertainty and $\frac{\Delta x}{x_0}$ the relative uncertainty (which we will briefly note δ_x). We can thus rewrite the expression of the efficiency in the light of this formulation:

$$\begin{aligned} \eta_{\min} &= \frac{P_{s\,\min}}{P_{e\,\max}} = \frac{C_{m\,\min}.\Omega_{m\,\min}}{U_{dc\pm\max}.I_{dc\,\max}} \\ &= \frac{C_{m0}.\Omega_{m0}\left(1 - \delta_{Cm}\right).\left(1 - \delta\Omega_m\right)}{U_{dc0}.I_{dc0}\left(1 - \delta_{U_{dc}}\right).\left(1 - \delta_{I_{dc}}\right)} \\ &= \eta_0 \frac{\left(1 - \delta_{Cm}\right).\left(1 - \delta\Omega_m\right)}{\left(1 + \delta_{U_{dc}}\right).\left(1 + \delta_{I_{dc}}\right)} \end{aligned} \quad \text{[A1.34]}$$

where η_0 is the nominal efficiency deduced directly from the measurements while supposing that they are exact. Analogously, we find the expression of the maximum efficiency $\eta_{\max}$:

$$\eta_{\min} = \frac{P_{s\,\max}}{P_{e\,\min}} = \eta_0 \frac{\left(1 + \delta_{Cm}\right).\left(1 + \delta_{\Omega_m}\right)}{\left(1 - \delta_{U_{dc}}\right).\left(1 - \delta_{dc}\right)} \quad \text{[A1.35]}$$

We thus obtain an uncertainty in the efficiency that we cal evaluate in an absolute manner as follows:

$$\Delta\eta = \eta_{\max} - \eta_{\min} = \eta_0 \left[\frac{\left(1 + \delta_{Cm}\right).\left(1 + \delta_{\Omega_m}\right)}{\left(1 - \delta_{U_{dc}}\right).\left(1 + \delta_{I_{dc}}\right)} \right. \quad \text{[A1.36]}$$
$$\left. - \frac{\left(1 - \delta_{Cm}\right).\left(1 - \delta_{\Omega_m}\right)}{\left(1 + \delta_{U_{dc}}\right).\left(1 + \delta_{I_{dc}}\right)} \right]$$

and then its simplification:

$$\Delta\eta = \eta_0$$

$$\times \left[\frac{\left(1+\delta_{C_m}\right)\cdot\left(1+\delta_{\Omega_m}\right)\cdot\left(1+\delta_{U_{dc}}\right)\cdot\left(1+\delta_{I_{dc}}\right) - \left(1-\delta_{C_m}\right)\cdot\left(1-\delta_{\Omega_m}\right)\cdot\left(1-\delta_{U_{dc}}\right)\cdot\left(1-\delta_{dc}\right)}{\left(1-\delta_{U_{dc}}\right)\cdot\left(1-\delta_{dc}\right)\cdot\left(1+\delta_{U_{dc}}\right)\cdot\left(1+\delta_{I_{dc}}\right)} \right]$$

$$[A1.37]$$

We can finally develop this expression and linearize by admitting that the relative uncertainties δ_x are small before 1. This stage is relatively monotonous and it can be replaced by "rough" evaluation of the variations of the following efficiency in each measurement. In our case, we obtain:

$$\begin{cases} P_{e\,\min} = 200 \times 0.9999 \times 50 \times 0.9999 = 9998.0001\,\text{W} \\ P_{e\,\max} = 200 \times 1.0001 \times 50 \times 1.0001 = 10002.0001\,\text{W} \\ P_{s\,\min} = C_{m\,\min}\Omega_{m\,\min} = 45 \times 0.999 \times 212 \times 0.999 \simeq 9520.9295\,\text{W} \\ P_{s\,\max} = C_{m\,\max}\Omega_{m\,\max} = 45 \times 1.001 \times 212 \times 1.001 \simeq 9559.0895\,\text{W} \end{cases}$$

$$[A1.38]$$

Consequently, we deduce:

$$\eta_{\min} = \frac{P_{s\,\min}}{P_{e\,\max}} \simeq 0.9519 \qquad\qquad [A1.39]$$

and:

$$\eta_{\max} = \frac{P_{s\,\max}}{P_{e\,\min}} \simeq 0.9561 \qquad\qquad [A1.40]$$

We then see that $\Delta\eta = \eta_{\max} - \eta_{\min}$ is 0.0042 (or 0.42 %): this result is significantly less good than the precision of the least performing devices (couple-meter and tachometer).

OBSERVATION A1.5.– We must, however, take note that this approach can turn out to be significantly more complex and dangerous in case we would wish to evaluate a more complex function. In this case, we will always prefer an approach based on mathematical tools.

A1.3.2. *Use of partial derivatives*

The use of partial derivatives therefore consists of calculating the different contributions of uncertainties. In fact, we have:

$$d\eta = \frac{\partial\eta}{\partial C_m}dC_m + \frac{\partial\eta}{\partial\Omega_m}d\Omega_m + \frac{\partial\eta}{\partial U_{dc}}dU_{dc} + \frac{\partial\eta}{\partial I_{dc}}dI_{dc} \quad [A1.41]$$

We note that these expressions of partial derivatives relative to C_m and Ω_m on the one hand and then U_{dc} and I_{dc} on the other hand are similar to the fact of the expression of the efficiency η. Effectively, we have:

$$\frac{\partial\eta}{dC_m} = \frac{\Omega_m}{U_{dc}.I_{dc}} \quad \text{et} \quad \frac{\partial\eta}{d\Omega_m} = \frac{C_m}{U_{dc}.I_{dc}} \quad [A1.42]$$

and then:

$$\frac{\partial\eta}{dU_{dc}} = -\frac{C_m.\Omega_m}{U_{dc}^2.I_{dc}} \quad \text{et} \quad \frac{\partial\eta}{dI_{dc}} = -\frac{C_m.\Omega_m}{U_{dc}.I_{dc}^2} \quad [A1.43]$$

We can then describe the link between the uncertainties (or more precisely their squares) according to the equation [A1.26] by considering the nominal values of different quantities (noted $U_{dc0} = 200\,\text{V}$, $I_{dc0} = 50\,\text{A}$, $C_m = 45\,\text{N.m}$ and $\Omega_m = 212\,\text{rad/s}$):

$$(\Delta\eta)^2 = \left(\frac{\Omega m}{U_{dc}.I_{dc}}\right)^2.(\Delta C_m)^2 + \left(\frac{C_m}{U_{dc}.I_{dc}}\right)^2.(\Delta\Omega_m)^2$$

$$+ \left(\frac{C_m.\Omega_m}{U_{dc}2.I_{dc}}\right)^2.(\Delta U_{dc})^2 + \left(\frac{C_m.\Omega_m}{U_{dc}.I_{dc}2}\right)^2.(\Delta I_{dc})^2 \quad [A1.44]$$

By putting:

$$\begin{cases} \Delta U_{dc}^{\smile} = 0.0002 \times U_{dc0} \\ \Delta I_{dc} = 0.0002 \times I_{dc0} \\ \Delta C_m = 0.002 \times C_{m0} \\ \Delta\Omega_m = 0.002 \times \Omega_{m0} \end{cases} \quad [A1.45]$$

We reach the following numerical result:

$$\Delta\eta' = 0.0027 \qquad\qquad [A1.46]$$

Given that the uncertainties take for the variables are the deviations between the minimal and maximal possible values, the magnitude obtained $\Delta\eta'$ is very coherent (0.27%) with the one obtained in the previous paragraph. We will note that this result is lower: not only is it an approached method but it also rests on one of the statistic hypotheses (in particular that of the independence of random variables). In fact, the result is probably more realistic than the one established with the intuitive approach. In fact, although this approach is prudent, it becomes the worst case when the worst case is not necessarily "probable". As an example, when we wish to size a bus or more generally means of public transport, we do not consider all of the passengers to be "overweight" and we consider the statistical "compensations" to better understand the real needs of such a vehicle.

A1.3.3. *Application of formulas*

In Table A1.2 we do not have a formula that is strictly adapted to our need. We apply formulas indicated "bit by bit" while not causing the quantities measured to appear (and uncertainties to appear only once).

It is therefore reasonable to calculate, at a first stage the uncertainty of the products (hence of the powers P_s and P_e) at the numerator and the denominator of the expression of efficiency η. We thus have:

$$\Delta P_e = U_{dc0}.\Delta I_{dc} + I_{dc0}.\Delta U_{dc} \qquad\qquad [A1.47]$$

and:

$$\Delta P_s = C_{m0}.\Delta\Omega_{m0} + \Omega_{m0}.\Delta C_{m0} \qquad\qquad [A1.48]$$

We can then carry out the numerical applications of the uncertainties and nominal values of the input and output powers:

$$\begin{cases} P_{e0} = U_{dc0}.I_{dc0} = 10\,000\,\text{W} \\ P_{s0} = C_{m0}.\Omega_{m0} = 9\,540\,\text{W} \\ \Delta P_e = 4\,\text{W} \\ \Delta P_s = 38.16\,\text{W} \end{cases} \qquad [\text{A1.49}]$$

And on this basis, we can use a formula of the table, this time relative to the division, to reach the final result on the uncertainty of the efficiency:

$$\Delta\eta" = \frac{P_{s0}\Delta P_e + P_{e0}\Delta P_s}{P_{e0}^2} = 0.0075 \qquad [\text{A1.50}]$$

This time we find a result that highly overestimates the efficiency uncertainties. In fact, we find an uncertainty of $0.75\,\%$ whereas by processing the worst case with the intuitive method[9] of the section A1.3.1, we had an uncertainty of $0.42\,\%$.

A1.3.4. *Overview*

The review of the methods presented is thus the following:

– the intuitive approach allows us *a priori* to obtain an uncertainty with a high degree of confidence, on the condition of evaluating the, or one of the worst case scenarios. We can mention that this can prove difficult in practice in the case of complex non-linear expressions;

– the use of these equations to the partial derivatives proves relatively simple and is applied to a wide range of mathematical expressions. On the contrary, there is an approached formula that does not allow us to estimate the

---

9 Which is strictly exact here given that we apply ourselves to a simple case.

uncertainty with a mastered degree of confidence (or at least easily). In exchange, the results often prove very satisfying in practice and take into account the independence of the variables[10];

– the application of the formulae is more restricted and it is also approached because we are obligated to process an expression bit by bit. In the case we have approached previously we reach an overestimation of the uncertainty and generally speaking, we can only evaluate the degree of confidence of the uncertainty calculated.

---

10 Which is often the case when we measure the different quantities with the separate and thus physically independent devices.

Appendix 2

Metric and Imperial Measuring Units

A2.1. Introduction

Measuring units are a source of confusion, errors and disagreement, particularly in electronics, since we mix the measuring units of the metric system and the Anglo-Saxon ones which are at the basis of the steps used for integrated circuit pins (not the standard of 2.54 mm but a tenth of inch). It is, therefore, useful to have conversion formulas or at least tables to make quick and error-free conversions. This is the goal of this annex that establishes several dimensional correspondences between lengths and surfaces by taking as reference the units powerly found in practice. For this, we will talk about:

– the lengths, especially the deviation (pitch) of the pins for electronic components (different standard values usually noted with thousandths of inches or "mil" and not in mm);

– the thicknesses in copper of the circuits printed, frequently noted in "ounces" (oz) and not in μm;

– the diameter of wires, noted in AWG (*American Wire Gauge*) and not in mm^2.

A2.2. Lengths, surfaces

The legal length unit in the international system is the meter. On the electronic scale, the usual unit is the millimeter. However, the majority of electronic components are characterized by deviations between pins that are sub-multiples of an inch. We can first quickly recall the measuring units of (power) lengths of the Anglo-Saxon system (the imperial system):

– inch: $25.4\,$mm;

– foot: $304.8\,$mm;

– mile: $1.609344\,$km.

In fact, in the field of electronics, only the inch is really used and we even work the most often with a sub-multiple noted "mil" for a thousandth inch. Consequently, a millimeter of an inch corresponds to $0.0254\,$mm whereas a millimeter corresponds to $39.37\,$mil.

The surfaces of printed circuits are often assessed by the manufacturers who are partially European (for pricing purposes) in "square inches" (in^2). We can easily check that $1\,in^2$ is equivalent to $\simeq\ 645.2\,mm^2$. Inversely, a square millimeter equals $1.55\ \times\ 10^{-3}\,in^2$ (or $1\,550$ mil^2).

A2.3. Standard deviations of printed circuit pins

The standard deviations of pins in *Dual Inline Package* type traversing integrated circuits as the one in Figure A2.1 have historically been calibrated on a tenth of inch (or $100\,$mil).

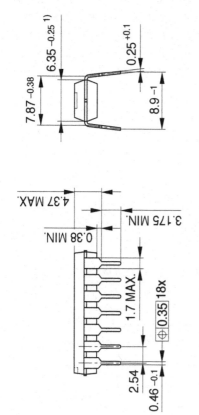

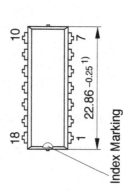

Index Marking

1) Does not include plastic or metal protrusion of 0.25 max. per side

Figure A2.1. *Example of DIP package with the given dimensions in inches (source: Infineon)*

Consequently, more compact packages have preserved this basic unit: we can mention small outline integrated circuit (SOIC) type for which the step is 50 mil (i.e. 1.27 mm). In contrast, the modern circuits with high pin density use ratings in millimeters, or micrometers (QFP control circuits – for *Quad Flat Package*, QFN – for *Quad Flat No-leads*, BGA – for *Ball Grid Array*). We can even see in Figure A2.2 an example of a bridge branch with FET transistors with gallium nitride (eGaN FET) developed by the EPC company whose packaging is of "BGA" type and whose step between ball is 400 μm for a nominal beads diameter of 200 μm.

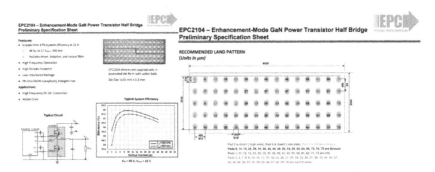

Figure A2.2. *Extracts from the datasheet of a bridge branch of FET GaN transistors*

A2.4. Copper mass and thickness of printed circuits

The copper thickness of the printed circuits are usually standardized[1]:

– the most usual ones are 35 μm[2];

---

1 Other thicknesses can be proposed by certain manufacturers (for instance, *Extreme Copper PCB* by EPEC Engineered Technologies – www.epectec.com)

2 Possibly after a stage of copper chemical deposit starting from an initial thickness of 17.5 to 18 μm. These low thicknesses are evidently unusable as well but this is less frequent.

– we have 70 μm for the "power" applications;

– finally, 105 μm when large currents must circulate in the small sized tracks (for instance, tracks of 2 mm allow 10 A to circulate with a heating limited to 20°C – see Chapter 6 of Volume 1 [PAT 15a]).

In the English-speaking world, the printed circuit boards are not characterized by the copper thickness but rather by the mass of copper layers over surface unit. More precisely, we note this value in "ounces/square foot" (oz/square foot). Let us recall that an ounce weighs 28.349 g in the *pounds* system (1/16 pounds) and by knowing the copper's volume mass (8.96 g/cm^3), we reach the following results:

– the 17.5 μm layers (or 0.7 mils) correspond to 0.5 oz/sq. foot (usually simply noted 0.5 oz);

– the 35 μm layers (or 1.4 mils) correspond to 1 oz;

– the 70 μm layers (or 2.8 mils) correspond to 2 oz;

– finally, the 105 μm layers (4.2 mils) correspond to 3 oz.

A2.5. Resistivity and strange units

Although this paragraph moves away from the general objective of this appendix, it emphasizes a unit that is seemingly surprising and that we find upon the detour of a technical notice such as [VIS 04], or a precision resistance datasheet: this is the resisitivity of a given thickness film. In fact, we usually find a characteristic of the resistive material in *ohm per square* (Ω/sq.) This unit is not defined according to space dimensions (length, surface or volume) but *uniquely on the basis of a geometrical form*: a square!

This seeming oddity is in fact connected to the two following hypotheses:

– a constant thickness e of the resistive layer;

– the coupling of the resistive "square" to the electric circuit on two opposing sides.

In these conditions, we can calculate very easily the resistance R_{sq} of this element by noting with a the length of the sides of the square by applying the formula:

$$R_{sq} = \rho \frac{l}{S} \qquad [A2.1]$$

where ρ is the resistivity of the material, l is the length of the resistive material and S its section. W can then note that for this geometry, we have:

$$\begin{cases} l = a \\ S = a.e \end{cases} \qquad [A2.2]$$

We then show that the resistance obtained does not depend on the length a:

$$R_{sq} = \frac{\rho}{e} \qquad [A2.3]$$

This result is particularly useful for characterizing the resistive films used in the resistances in "thin layers" but also the semiconductors (measure of resistivity upon wafer). Besides, this could affect the design of the geometry of certain shunts as the one in Figure A2.3. We can see there certain areas that are clearly made of square blocks in series and parallel, possibly coupled by a laser beam (this is a precision resistances network used in the Keithley DMM7510 multimeter – *Laser Trimmed Precision Thin-Film Resistor Network*).

A2.6. Wire diameters

In the area of wires, we are generally interested in their section because it allows us to determine the maximum power that we can circulate once the maximum authorized density has been defined (for instance $5\,A/mm^2$ for "thermally confined" copper and so in a situation of low heat sinking). In France, different standardized sections (in mm^2) of copper wires are available. These wires can be made of several strands[3], subtle or rigid made of a single conductor. The given section is obviously that of copper and not of the whole wire (including the insulating sheath).

Figure A2.3. *Precision resistances network used in a Keithley DMM7510 multimeter (source: Wikipedia, Binarysequence)*

We thus find power sections such as: $0.75\,mm^2$, $1.5\,mm^2$, $2.5\,mm^2$, $4\,mm^2$ and $6\,mm^2$. In practice, the density of the power of $5\,A/mm^2$ is very conservatory (i.e. prudent) because the wires of $1.5\,mm^2$ are used for power sockets $16\,A$ in household installations. The rule applied in this case is then to consider a power density of $10\,A/mm^2$. The power density is evidently connected to the implementation of the electric circuit: is the cabling dense and could the heat be easily evacuated. Generally speaking, household appliances are not particularly dense and the thermal aspect of cables is therefore a major problem but this is not always the case.

---

3 These are subtle wires with non-varnished conductors. It is not the Litz wire presented in Chapter 2 of Volume 3 [PAT 15c] whose interest is to minimize the skin effect at high frequency.

Once more, Anglo-Saxon notations are different for cable description. The measuring unit used in the United States is the *American Wire Gauge* (AWG) which was updated by J.R. Brown in 1857. The numerical value indicated for a cable is even bigger as the conductor is of a small diameter. The AWG number allows us to obtain the diameter d (in mm) and the section S of the wire (in mm^2) starting from the following formulas:

$$d\,[\text{mm}] = 0.127 \times 92^{\frac{36-n}{39}} \qquad\qquad [\text{A2.4}]$$

and:

$$S\,[\text{mm}^2] = \frac{\pi}{4}d^2 = 0.012668 \times 92^{\frac{36-n}{19,5}} \qquad\qquad [\text{A2.5}]$$

Index n	AWG	Diameter (in mm)	Section (in mm^2)
−3	0000 or 4/0	11,7	107
−2	000 or 3/0	10,4	85,0
−1	00 or 2/0	9,26	67,4
0	0 or 1/0	8,25	53,5
1	1	7,35	42,4
2	2	6,54	33,6
3	3	5,83	26,7
4	4	5,19	21,1
5	5	4,62	16,8
6	6	4,11	13,3
7	7	3,66	10,5
8	8	3,26	8,36
9	9	2,91	6,63
10	10	2,59	5,26
11	11	2,30	4,17
12	12	2,05	3,31
13	13	1,83	2,62
14	14	1,63	2,08
15	15	1,45	1,65
16	16	1,29	1,31
17	17	1,15	1,04
18	18	1,02	0,823

Table A2.1. *AWG, diameters and sections of wire*

We can see in Table A2.1 several characteristics of the wires for n varying from -3 to 18. The null and negative values are then noted as follows:

– for $n = 0$, we note 0 AWG;

– for $n = -1$, we note 00 AWG or 1/0 AWG;

– for $n = -2$, we note 000 AWG or 2/0 AWG;

– etc.

Errata

Although all five volumes of the "Electronics Applied to Industrial Systems and Transports" publication have been thoroughly reviewed, some typos remain. Below is a list of errors identified in the first four volumes as of January 2016. Feedback can be sent to the author by email at nicolas.patin@ieee.org. The error list will be updated on the basis of the feedback received.

Volume 1 [PAT 15a]: Synthetic Methodology to Converters and Components Technology

– Chapter 4 (page 86): There are two types of polarized capacitors: the first bullet point mentions electrolytic *aluminum* capacitors. However, the electrodes' nature (aluminum and its oxide) is omitted even though tantalum capacitors are also of electrolytic nature (this creates an ambiguity).

Volume 3 [PAT 15c]: Switching Power Supplies

– Chapter 5 (page 76): item (2nd bullet point on the page).

Bibliography

[BOU 12] BOUDELLAL, M., *La pile combustible. L'hydrogène et ses applications*, 2nd ed., Dunod, Paris, 2012.

[BOU 08] ARDLEY, T., *First Principles of a Gas Discharge Tube (GDT) Primary Protector*, available at: www.bourns.com, 2008.

[CAS 07] CATELVECCHI D., *"Spinning into control– High-tech reincarnation of an ancient way of storing energy"*, Science News, vol. 171, no. 20, p. 312, 2007.

[COU 14] COUSSEAU R., PATIN N., MONMASSON E. *et al.*, "Advanced electric model of aluminum electrolytic conductors with diffusive element," 11th *International Conference on Modeling and Simulations of Electric Machines, Converters and Systems*, Valence, pp. 19–22, May 2014.

[DAH 96] DAHONO P.A., SATO Y., KATAOKA T., "Analysis and minimization of ripple components of input power and voltage of PWM inverters," *IEEE Transactions on Industry Applications*, vol. 32, no. 4, pp. 945–950, July 1996.

[DIG 15] DIGIKEY, *Application note for Inrush power Limiters (ICL)*, available at: www.digikey.com, 2015.

[EME 11] EMERSON NETWORK POWER, *Surge protection reference guide*, available at: www.emersonnetworkpower.com, 2011.

[EPC 13] EPCOS, *NTC Thermistors for Inrush power Limiting*, Application Note, available at: en.tdk.eu, 2013.

[EPS 15] EPSIC, *Cours en ligne d'électrotechnique, fonctionnement des instruments de mesure*, available at: http://www.epsic.ch/cours/electrotechnique/ theorie/instrufonct/ 240.html, 2015.

[FER 06] FERRAZ-SHAWMUT, *Power Semiconductor Fuse Applications Guide*, available at: www.europowercomponents.com, 2006.

[GAG 14] GAGNEUR L., KVIESKA P., MERIENNE L. *et al.*, Structure for modulating the voltage of a battery and the active equilibration thereof, Patent WO 2014057192 A2, Renault SAS, 2014.

[GAS 05] GASPERI M., "Life prediction modelling of bus conductors in AC variable frequency drives", *Conference Records of 2005 Annual Pulp and Paper Industry Technical Conference*, Jacksonville, FL, pp. 141–146, 20–23 June, 2005.

[GER 97] GERL M., ISSI J.-P., *Physique des matriaux*, vol. 8, Presses Polytechniques Universitaires Romandes, Lausanne, 1997.

[GRA 98] GRACIET M., PINEL J., *Protection contre les perturbations – Composants de protection*, available at: www.techniques-ingenieur, 1998.

[HEM 06] HEMERY C.-V., Etudes des phenomènes thermiques dans les batteries Li-ion, PhD Thesis, University of Grenoble, 2006.

[HOB 05] HOBRAICHE J., *Comparaison de stratégies de modulation largeur d'impulsion triphasées – Application l'Alterno-Démarreur*, Université de Technologie de Compiègne, 2005.

[INE 06] INERIS, *Guide ATEX pour les néophytes*, available at: www.ineris.fr, 2006.

[INR 12] INRS, *Charge des batteries d'accumulateurs au plomb*, available at: www.inrs.fr, 2012.

[INS 15] INSTALLATIONS ELECTRONIQUES, Appareils électriques – l'électroménager, available at: www.installations-electriques.net/Apelm/Corpus_techmes.htm, 2015.

[KEI 10] KEITHLEY, *Low Level Measurement Handbook*, 6th ed., available at http://info.tek.com/KI-Low-Level-Measurements-Handbook-LP.html, USA, 2010.

[LAC 13] SAMANTHA LACROIX, Modélisation et commande d'une chaîne de conversion pour véhicule électrique intégrant la fonction de charge des batteries, PhD Thesis, Université Paris-Sud, 2013.

[MAT 01] MATHIEU H., *Physique des semi-conducteurs et des composants électroniques*, 5th ed., Dunod, Paris, 2001.

[MER 13] MERLET C., Modélisation de l'adsorption des ions dans les carbones nanoporeux, PhD Thesis, Université Pierre et Marie Curie, Paris, 2013.

[MOT 95] MITTER C.S., *Active Inrush power Limiting Using MOSFET*, Application Note AN1542, Motorola Inc., 1995.

[NGU 11] NGUYEN T.D., Etude de stratégies de modulation de convertisseurs statiques dédiés la réduction des perturbations conduites en environnement embarqué, PhD Thesis, Université de Technologie de Compiègne, 2011.

[PAT 10] PATIN N., NGUYEN T.D., FRIEDRICH G., "Analyse des sollicitations sur les condensateurs de filtrage du bus continu d'onduleur pour applications embarquées en fonction de la stratégie de modulation", *Conference EPF'2010*, Saint-Nazaire, 30 June – 2 July 2010.

[PAT 15a] PATIN N., *Power Electronics Applied to Industrial Systems and Transport*, vol. 1, ISTE Ltd., London and Elsevier, Oxford, 2015.

[PAT 15b] PATIN N., *Power Electronics Applied to Industrial Systems and Transport*, vol. 2, ISTE Ltd., London and Elsevier, Oxford, 2015.

[PAT 15c] PATIN N., *Power Electronics Applied to Industrial Systems and Transport*, vol. 3, ISTE Ltd., London and Elsevier, Oxford, 2015.

[PAT 15d] PATIN N., *Power Electronics Applied to Industrial Systems and Transport*, vol. 4, ISTE Ltd., London and Elsevier, Oxford, 2015.

[RAD 04] RADIOMETER ANALYTICAL, *Conductivity Theory and Practice*, available at: www.analytical-chemistry.voc.gr, 2004.

[RAG 68] RAGONE D.V., *Review of Battery Systems for Electrically Powered Vehicles*, SAE Technical Paper 680453, 1968.

[REV 14] REVANKAR S.T., MAJUMDAR P., *Fuel Cells: Principles, Design, and Analysis*, CRC Press, New York, 2014.

[ROG 12] WALTER ROGOWSKI and STEINHAUS W., "Die Messung der magnetischen Spannung," *Archiv für Elektrotechnik*, vol. 1, no. 4, pp. 141–150, 1912.

[ROS 11] ROSCHER M.A., BOHLEN O., VETTER J., "OCV hysteresis in li-ion batteries including two-phase transition materials," *International Journal of Electrochemistry*, vol. 2011, pp. 1–6, 2011.

[SCH 11] SCHURTER, *Guide to fuse selection*, available at: www.schurter.ch, 2011.

[SEM 14] ZURNACI M., *MiniSKiiP Generation II, Technical explainations*, available at: www.semikrom.com, 2014.

[SIE 04] TARZWELL R., BAHL K., *High Voltage Printed Circuit Design & Manufacturing Notebook*, Sierra Proto Express, 2004.

[SIM 13] SIMON P., GOGOTSI Y., "Capacitive energy storage in nanostructured carbon electrolyte systems," *Accounts of Chemical Research*, vol. 46, no. 5, pp. 1094–1103, 2013.

[SOC 04] STEINER M., *Catalogue général SOCOMEC*, available at: www.socomec.com, 2004.

[VIE 03] VIELSTICH W., LAMM A., GASTEIGER H.A., *Handbook of Fuel Cells: Fundamentals, Technology, Applications*, Wiley, New York, 2003.

[VIS 04] VISHAY MIC Technology – *Applications and Design of Thin Film Resistors*, Vishay Electro-Films, 2004.

[VOG 11] VOGELSBERGER M.A., WIESINGER T., ERTL H., "Life-cycle monitoring and voltage-managing unit for DC-link electrolytic conductors in pWM converters," *IEEE Transactions on Power Electronics*, vol. 26, no. 2, pp. 493–503, February 2011.

[XAP 05] ALEXANDER M., *Power Distribution System (PDS) Design: Using Bypass/Decoupling conductors*, Application Note, XAPP623, available at: www.xilinx.com, February 2005.

Index